Mathematical Problems in Linear Viscoelasticity

SIAM Studies in Applied Mathematics

This series of monographs focuses on mathematics and its applications to problems of current concern to industry, government, and society. These monographs will be of interest to applied mathematicians, numerical analysts, statisticians, engineers and scientists who have an active need to learn useful methodology.

Series List

Vol. 1. *Lie-Bäcklund Transformations in Applications*
Robert L. Anderson and Nail H. Ibragimov

Vol. 2. *Methods and Applications of Interval Analysis*
Ramon E. Moore

Vol. 3. *Ill-Posed Problems for Integrodifferential Equations in Mechanics and Electromagnetic Theory*
Frederick Bloom

Vol. 4. *Solitons and the Inverse Scattering Transform*
Mark J. Ablowitz and Harvey Segur

Vol. 5. *Fourier Analysis of Numerical Approximations of Hyperbolic Equations*
Robert Vichnevetsky and John B. Bowles

Vol. 6. *Numerical Solution of Elliptic Problems*
Garrett Birkhoff and Robert E. Lynch

Vol. 7. *Analytical and Numerical Methods for Volterra Equations*
Peter Linz

Vol. 8. *Contact Problems in Elasticity: A Study of Variational Inequalities and Finite Element Methods*
N. Kikuchi and J. T. Oden

Vol. 9. *Augmented Lagrangian and Operator-Splitting Methods in Nonlinear Mechanics*
Roland Glowinski and P. Le Tallec

Vol. 10. *Boundary Stabilization of Thin Plates*
John E. Lagnese

Vol. 11. *Electro-Diffusion of Ions*
Isaak Rubinstein

Vol. 12. *Mathematical Problems in Linear Viscoelasticity*
Mauro Fabrizio and Angelo Morro

Mauro Fabrizio
Angelo Morro

Mathematical Problems in Linear Viscoelasticity

Society for Industrial and Applied Mathematics
siam® *Philadelphia / 1992*

Library of Congress Cataloging-in-Publication Data

Fabrizio, Mauro, 1940–
Mathematical problems in linear viscoelasticity / Mauro Fabrizio and Angelo Morro.
p. cm. – (SIAM studies in applied mathematics ; vol. 12)
Includes bibliographical references and index.
ISBN 0-89871-266-1
1. Viscoelasticity. I. Morro, Angelo, 1946– II. Title.
III. Series: SIAM studies in applied mathematics ; 12.
QA931.F33 1992
531'.382–dc20 92-4050

Contents

To the memory of D. Graffi and V. Volterra,
pioneers in the theory of systems with memory

Preface

The aim of this book is to investigate the connection between thermodynamic restrictions and the well posedness of initial and boundary value problems. On the one hand, we elaborate a thorough thermodynamic analysis of linear viscoelasticity; new results are established and previous ones are shown to follow as particular cases from the general scheme. On the other hand, we show that significant improvements can be obtained in existence, uniqueness, and asymptotic stability theorems by starting from the thermodynamic restrictions as mathematical hypotheses for the initial-boundary value problems. On the whole, various subjects are examined: well posedness of Cauchy's problems, variational and extremum principles, wave propagation, thermodynamics of continuous media, modelling of viscoelastic materials. The central role played by thermodynamic restrictions on the properties of the solutions makes these subjects deeply interrelated.

All along, mathematical statements and detailed proofs of results are framed in the pertinent physical context. The material developed is essentially self-contained. The reader is required to have general familiarity with standard techniques of modern analysis and basic concepts in continuum thermodynamics. This book is then of interest to mathematicians, physicists, and engineers.

The authors express their gratitude to Professors J. A. Nohel and M. Slemrod for the opportunity to write this book. To Professor C. Giorgi they are indebted for invaluble discussions and criticism. They also wish to thank Professors G. Caviglia and B. Lazzari and Doctors G. Gentili and E. Vuk for helpful suggestions.

Mauro Fabrizio
University of Bologna, Italy

Angelo Morro
University of Genova, Italy

Introduction

The material behaviour of solids and fluids is described through various models which constitute the subject of corresponding chapters of continuum mechanics. Elasticity is the simplest model and is applicable whenever instantaneous response of the material in a purely mechanical framework is an adequate approximation. Very often, though, the behaviour of the material is not merely elastic. This is because the material exhibits hereditary properties and, simultaneously, dissipative effects. Apart from the peculiar hysteretic phenomena, hereditary properties are well described by viscoelasticity where, whether or not thermal effects are considered, the mechanical response of the material is taken to be influenced by the previous behaviour of the material itself. It then should come as no surprise that, along with the progress of continuum mechanics, the literature has devoted an increasing attention to viscoelasticity.

Among the recent advances of continuum mechanics of remarkable importance is the theory of materials with fading memory which is by now well established in various monographs. More specifically, we have in mind the theory (or the theories) developed within the context of rational thermodynamics where, roughly speaking, the second law of thermodynamics is viewed as the basic tool for an a priori characterization of the description of the material behaviour. Of course, the idea of casting viscoelasticity within the thermodynamics of materials with fading memory is not new at all. Indeed, we can say that this task has been performed for more than twenty years. Nevertheless, recent improvements and results about initial and boundary value problems in linear viscoelasticity suggest that an exhaustive analysis of the thermodynamic theory of linear viscoelasticity can become a useful reference for future developments.

The connection between thermodynamic restrictions and properties of the solution is stronger than it might seem at first sight. Recently we have reviewed carefully the thermodynamic theory of viscoelasticity and have established new results. Next, in conjunction with some collaborators, we have investigated the role of the new, or more precise, thermodynamic restrictions within the context of mathematical problems in the dynamics of viscoelastic bodies. In particular, we have devoted attention to the role of the thermodynamic restrictions in connection with the well posedness of appropriate differential problems. The promising results obtained, along with the need of an up-to-date mathematical theory of viscoelasticity, have led us to write the present book.

Our aim is to emphasize how the restrictions placed by thermodynamics on the constitutive functionals, chiefly Graffi's inequality, are of fundamental importance when investigating properties of the solution to pertinent problems in viscoelasticity. In both statics and dynamics, the thermodynamic restrictions lead to the well posedness of initial-boundary value problems, yield the minimum of energy-like functionals, and ensure the propagation condition and the decay of waves. By means of counterexamples, the thermodynamic restrictions are shown to be not only sufficient but also necessary for existence, uniqueness, minimum properties, and so on. As is apparent at a glance, the material in this book has been devised and arranged so as to show how the principles of thermodynamics provide exhaustive physical assumptions since they result in a satisfactory, mathematical modelling of problems in linear viscoelasticity.

In view of the valuable monographs on viscoelasticity which have appeared even in recent times, we have chosen not to develop the whole theory from scratch. The preliminaries exhibited in this book are motivated only by our attempt to write a self-contained work. In our view, the style renders the procedure mathematically precise but does not overburden the text with cumbersome notation. Rather, we have chosen to give reasonable space to the various developments especially in connection with topics still under speculation or not yet fully established.

This book is divided into seven chapters whose content may be summarized as follows. Since we have chosen to model viscoelasticity within the theory of materials with fading memory, the first chapter is devoted to an outline of concepts, definitions, and properties which are by now customary in such a theory. In particular, we review the definition and main properties of temporal histories and the principle of fading memory for the response functional(s) of the material. Then the response functional is given a fading memory space of histories as domain and is supposed to allow the application of the chain rule. Finally, the essentials of continuum mechanics are outlined for the purpose of gathering the fundamental notions applied in this book.

The second chapter provides a general scheme of the thermodynamics of simple materials (wherein we cast viscoelastic solids and fluids). We begin with the notion of thermodynamic states and processes and then we examine the statements for the first and second law. A version of the second law is given for approximate cycles which are usually in order when dealing with materials with memory. Then we consider reversible processes, a topic which, in our mind, deserves more attention in the study of continuum thermodynamics. Once reversible processes are characterized, we show that processes in viscoelasticity are not reversible, which motivates the next statement of the second law of thermodynamics where reversibility enters in a precise way. The connection to more familiar versions of the second law or the dissipation principle is examined by considering the statement of dissipativity, the Clausius-Duhem inequality, and the Clausius inequality.

The third chapter develops the modelling assumptions of viscoelastic solids and fluids along with a detailed, thorough derivation of thermodynamic restrictions. In conjunction with cyclic processes, a set of restrictions on the relaxation function is derived which proves to be also sufficient for the validity of the second law in the general case of approximate cycles. As an aside, the dissipativity condition (for viscoelastic solids) is shown to be equivalent to the second law along with the positive definiteness of the equilibrium elastic modulus. Then the determination of the free energy functional is investigated which, according to thermodynamics, is the potential for the viscoelastic stress functional. It is shown that in general this potential is nonunique and how Volterra's and Day's are the most natural forms. This problem finds an interesting formulation within a suitable finite-dimensional context of thermodynamics with internal variables.

The fourth chapter deals with the properties of the solution to various problems in viscoelasticity. In connection with the quasi-static problem, some remarkable counterexamples to existence and uniqueness are revisited, which show how the function space ensuring well posedness is related to the chosen relaxation function. Existence and uniqueness are proved for the quasi-static problem when the body force is time-harmonic. The fully dynamic problem is then considered and existence, uniqueness, and stability for Cauchy's problem, along with the asymptotic behaviour, are investigated in detail.

The fifth chapter consists of two parts. First, convolution-like variational formulations are determined in various descriptions (relaxation and creep type) by following the theory pertaining to the inverse problem of the calculus of variations. Then we devote attention to extremum principles. Essentially minimum properties are established within various frameworks: by looking at functionals which involve a weight function in the domain of the Laplace transform (of the solution) or in a class of separable displacement fields. Moreover, the connection of thermodynamics with the convexity of a suitable functional is settled. Then viscoelastic fluids are considered and extremum principles are established for both incompressible and compressible fluids.

The sixth chapter reviews and generalizes some results on wave propagation and, again, the strict connection between thermodynamic restrictions and properties of the solution is shown. This is performed first in the case of propagation through singular surfaces. Shock waves, waves of higher order, and induced discontinuities are shown to decay as a consequence of the thermodynamic restrictions. In this conjunction, the possibility of modelling energetic materials still compatible with thermodynamics is investigated by having recourse to materials with internal variables. Next time-harmonic waves are considered with special emphasis on the inhomogeneity property which is typical of dissipative bodies like the viscoelastic ones.

The last chapter is devoted to the Rayleigh problem in viscoelastic solids and fluids with a relaxation function which is unbounded or has an unbounded initial derivative. Such unboundedness, which seems to be motivated by microscopic considerations or experimental observations on polymeric fluids, leads to interesting mathematical problems. Upon reexamining the main thermodynamic aspects for unbounded relaxation functions, we show how, once again, the thermodynamic restrictions turn out to be sufficient for determining the solution to initial-boundary value problems such as the Rayleigh problem for both solids and fluids.

As a general comment regarding the content of this book, we observe that in any mathematical question, such as existence, uniqueness, and stability of solutions to Cauchy's problems, extremum properties, and wave propagation and amplitude evolutions, thermodynamic restrictions prove to be the right or natural way to get well-posed problems. Indeed, in connection with problems already investigated at length in the literature, thermodynamic restrictions have provided models of viscoelastic materials which simplify the mathematical analysis while preserving the qualitative results derived through more involved assumptions.

On the whole, various subjects are investigated: well posedness of Cauchy's problems, variational principles, wave propagation, thermodynamics of continuous media, modelling of viscoelastic materials. What is more, these subjects are shown to be deeply interrelated. All along, mathematical statements and detailed proofs of results are framed in the pertinent physical context. This book is then likely to be of interest to mathematicians, physicists, and engineers.

Chapter 1

Preliminaries on Materials with Fading Memory

1.1. Notation. The space under consideration is the three-dimensional Euclidean point space $\mathcal{E}^3$. Vectors are elements of the associated translation space V and are denoted by boldface minuscules. Second-order tensors are meant as linear transformations of V into V. Sym is the set of symmetric (second-order) tensors while sym denotes the symmetric part of a tensor. Lin stands for the set of all tensors; Lin(Sym) is the set of all linear transformations of Sym into Sym; Lin^+ is the subset of Lin whose elements have a positive determinant; Psym is the set of elements of Sym which are positive definite; Orth^+ is the set of all rotations. Boldface majuscules usually denote elements of Lin or Lin(Sym); $\mathbf{1}$ is the identity of Lin. The symbols tr and det stand for the trace and the determinant of tensors while the superscript T, e.g., $\mathbf{M}^T$, denotes the transpose. Letting $\mathbf{u}, \mathbf{v} \in V$, we write $\mathbf{u} \cdot \mathbf{v}$ for the standard inner product. If $\mathbf{L}, \mathbf{M} \in \text{Lin}$, then $\mathbf{L} \cdot \mathbf{M}$ stands for $\text{tr}(\mathbf{L}\mathbf{M}^T)$. If $\mathbf{M}$ is an element of Lin or Lin(Sym), then the writing $\mathbf{M} > 0$ (< 0) means that $\mathbf{M}$ is positive (negative) definite in V or in Sym. For any two vectors $\mathbf{u}, \mathbf{v}$, $\mathbf{u} \otimes \mathbf{v}$ denotes the tensor product.

Usually Ω is a domain in $\mathcal{E}^3$. $C^n(\Omega)$ is the linear space whose elements are real functions having continuous partial derivatives up to the order n in Ω; $C_0(\Omega)$ is the linear space of continuous functions with compact support in Ω and $C_0^n(\Omega) = C_0(\Omega) \cap C^n(\Omega)$. The symbol $W^{l,p}(\Omega)$ stands for the spaces of functions which, along with their partial derivatives up to the order l, are in the Banach space $L^p(\Omega)$. For brevity we let $H^l = W^{l,2}$. Further, $W_0^{l,p}(\Omega)$ is the completion of $C_0^\infty(\Omega)$, namely the set of C^∞-functions with compact support in $W^{l,p}(\Omega)$ for $p \in [1, \infty)$. $W^{-l,q}(\Omega)$ is the dual space of $W_0^{l,p}(\Omega)$ via the inner product on $L^2(\Omega)$ for $p \in (1, \infty)$; $H^{-l}(\Omega) = W^{-l,2}(\Omega)$. Particular spaces are defined when introduced.

A superposed dot denotes the (material) time derivative while $\nabla, \nabla\cdot$, and

$\nabla\times$ stand for the gradient, the divergence, and the curl, respectively. Moreover, $\mathbb{R}, \mathbb{R}^+, \mathbb{R}^{++}, \mathbb{R}^-$ are the set of reals, positive reals, strictly positive reals, and negative reals, respectively. With regard to the set $\mathbb{C}$ of complex numbers, $\Re$ and $\Im$ denote the real and the imaginary parts, respectively; $\mathbb{C}^+$ and $\mathbb{C}^{++}$ stand for the half planes with positive and strictly positive real part, respectively. Usually the complex variable is denoted by p and α, β are the real and imaginary parts, respectively.

For any function $g \in L^1(\mathbb{R}^+)$, $g_L(p)$ denotes the Laplace transform at $p \in \mathbb{C}^+$. If $g \in L^1(\mathbb{R})$ we let g_F be its Fourier transform on $\mathbb{R}$. Further, g_c and g_s denote the half-range Fourier cosine and sine transform while g_C and g_S denote the full-range analogues. In connection with differentiable functionals on appropriate function spaces, we let $d\Psi(u|h)$ be the first-order differential of Ψ at u in the direction h, namely $\Psi(u+h) - \Psi(u) = d\Psi(u|h) + o(\|h\|)$. Similarly, $d^2\Psi(u|h)$ denotes the second-order differential such that $\Psi(u+h) - \Psi(u) = d\Psi(u|h) + \frac{1}{2}d^2\Psi(u|h) + o(\|h\|^2)$.

1.2. Histories and fading memory. Let W denote a finite-dimensional vector space (possibly V, Lin) and consider functions which map $\mathbb{R}$ into W. A *history* is a function defined on $\mathbb{R}^+$, with values in W, while a *past history* is a function defined on $\mathbb{R}^{++}$, with values in W. Given a function ϕ on $\mathbb{R}$ and a time $t \in \mathbb{R}$, we define the history ϕ^t by

$$\phi^t(s) = \phi(t-s), \qquad s \in \mathbb{R}^+.$$

The function ϕ^t is called the *history of ϕ up to time t*. The restriction of ϕ^t to $\mathbb{R}^{++}$ is called the *past history of ϕ up to time t* and denoted by $_r\phi^t$. For any history ϕ^t,

$$\phi^t(0) = \phi(t)$$

is called the *present value* of ϕ^t. The history ϕ^t such that $\phi^t(s) = 0$ for every s in $\mathbb{R}^+$ is called the *zero history.* The *constant history* $\bar{\phi}^\dagger$ is defined as $\phi^t(s) = \bar{\phi}, s \in \mathbb{R}^+$. For any history ϕ^t and $\tau > 0$ we define the *static continuation* $\phi^{t(\tau)}$ of ϕ^t, by the amount τ, as the history

$$\phi^{t(\tau)}(s) = \begin{cases} \phi^t(0), & s \in [0,\tau), \\ \phi^t(s-\tau), & s \in [\tau,\infty). \end{cases}$$

Similarly, we define the *τ-section* $\phi^t_{(\tau)}$ of ϕ^t as

$$\phi^t_{(\tau)}(s) = \phi^t(s+\tau), \qquad s \in [0,\infty).$$

Since, by definition,

$$\phi^t_{(\tau)}(s) = \phi^{t-\tau}(s),$$

then $\phi^t_{(\tau)}$ is the history up to the earlier time $t - \tau$. The *retardation* ϕ^t_γ, $\gamma \in (0,1)$, of ϕ^t is defined as

$$\phi^t_\gamma(s) = \phi^t(\gamma s), \qquad s \in [0, \infty).$$

Let Ω be a domain (open set) in the three-dimensional Euclidean point space $\mathcal{E}^3$. A *field* is a function $\phi : \Omega \to W$; a function $\phi : \Omega \times \mathbb{R} \to W$ is a *time-dependent field.* For any $\mathbf{x} \in \Omega$ the symbol $\phi^t(\mathbf{x}, \cdot)$ denotes the history of ϕ at $\mathbf{x}$ up to time t. To save writing, we omit the dependence on $\mathbf{x}$ whenever no ambiguity arises.

Roughly speaking, a system with memory is such that the output (or response), u say, at time $t \in \mathbb{R}$, is determined by a functional $\mathcal{F}$ on a suitable set of histories ϕ^t up to time t,

$$u(t) = \mathcal{F}(\phi^t). \tag{1.2.1}$$

In practice, the entire history of a quantity ϕ in a system can never be known. Then an equation like (1.2.1) has an operative meaning only if additional assumptions are made. The main assumption is that the memory is fading; recent values of $\phi^t(s)$, s around 0^+, rather than the remote values, $s \to \infty$, have the main effect on the output u. While the idea is quite intuitive, to make it operative we need a precise mathematical statement. We first outline the theory of fading memory as elaborated by Coleman and Noll [25], [26].

In essence, we consider the domain of $\mathcal{F}$ with possible restrictions on ϕ. We make a choice of topology on this domain such that two histories are close if they are close in the recent past. Then we assume the functional $\mathcal{F}$ to be smooth in this topology, which amounts to letting $\mathcal{F}$ weakly dependent on remote values of ϕ. This scheme may be realized as follows.

Let $k : \mathbb{R}^+ \to \mathbb{R}^{++}, k(0) = 1$, be a function which decays monotonically to zero for large s in such a way that, when $\alpha > 1, \lim s^\alpha k(s) = 0$, as $s \to \infty$. A function satisfying these conditions is called an *influence function* of order α (or obliviator) and, as we shall see, characterizes the rate at which the memory fades. To specify the smoothness of $\mathcal{F}$, we define a $L^2(\mathbb{R}^+)$ norm of ϕ^t, with weight k, as the "recollection"

$$\|\phi^t\|_k = \left(\int_0^\infty |\phi^t(s)|^2 k(s)\, ds \right)^{1/2} < \infty \tag{1.2.2}$$

where

$$|\phi^t(s)| = [\phi^t(s) \cdot \phi^t(s)]^{1/2},$$

the symbol "$\cdot$" denoting the inner product in the vector space W. The collection of all histories with finite norm (1.2.2) forms a Banach space. Indeed, the natural definition of inner product

$$\langle \phi^t_1, \phi^t_2 \rangle_k = \int_0^\infty \phi^t_1(s) \cdot \phi^t_2(s)\, k(s)\, ds$$

extends the Banach space to a Hilbert space, usually denoted by S_k. We can now state the following principle.

Weak principle of fading memory. *There exists an influence function k, of order $\alpha > 1$, such that the response functional* (1.2.1) *is defined and continuous for histories ϕ^t in a neighbourhood of the zero history in the space S_k.*

Accordingly, it is the continuity of $\mathcal{F}$, in terms of the topology defined by (1.2.2), that expresses the assumption of fading memory. Since k is a decreasing function, at least for large values of s, two histories differ little in norm if their values are close to each other in the recent past (small s), no matter how they differ in the distant past.

Of course, for a given response functional the choice of k is not unique. The principle requires only that at least one such influence function exists.

More severe statements of fading memory involve stronger smoothness assumptions on the response functional. Letting n be a positive integer, typically $n = 1, 2$, the following principle is often assumed.[1]

Strong principle of fading memory. *There exists an influence function k, of order greater than $n + 1$, such that the response functional* (1.2.1) *is defined and n-times Fréchet-differentiable in a neighbourhood of the zero history in the space S_k.*

This is the essence of what is sometimes called the elementary theory [19], [20] of fading memory. Though based on the same spirit, more sophisticated versions have been given by Coleman, Mizel, Owen, and Wang [23], [24], [29], [130]. In such versions the properties of k are taken as follows.

A real-valued, Lebesgue-measurable function k on $\mathbb{R}^{++}$ is called an influence function with the relaxation property if

(i) $k(s) > 0$ almost everywhere on $\mathbb{R}^{++}$ and $k \in L^1(\mathbb{R}^+)$;

(ii) the functions $\bar{K}$ and $\underline{K}$, defined by

$$\bar{K}(\tau) := \operatorname*{ess.sup}_{s \in \mathbb{R}^{++}} \frac{k(s+\tau)}{k(s)},$$

$$\underline{K}(\tau) := \operatorname*{ess.sup}_{s \in \mathbb{R}^{++}} \frac{k(s)}{k(s+\tau)},$$

have finite values on $\mathbb{R}^+$;

(iii)

$$\sup_{\tau \in \mathbb{R}^+} \bar{K}(\tau) < \infty.$$

[1]The strong principle of fading memory may be overly restrictive and, in fact, models in rheology, such as the BKZ fluid [4], do not comply with it. The role of this principle, though, may be appreciated in nonlinear theories.

At least one influence function is supposed to exist and then the recollection is taken as the L^p_k norm, $p \in [1, \infty)$, namely

$$\|\phi^t\|^p_k = \int_0^\infty |\phi^t(s)|^p k(s)\, ds.$$

As shown in [24], if a real-valued function k on $\mathbb{R}^{++}$ has properties (i) and (ii), then there exist positive numbers a, b, and c such that

$$a \exp(-bs) < k(s) < c \quad \text{a.e. on} \quad \mathbb{R}^{++},$$

and, moreover, $sk(s) \to 0$, essentially, as $s \to \infty$. Moreover, if k has properties (i) and (ii), then the monotone decreasing property of k is sufficient, but not necessary, for k to meet (iii).

It is natural to ask whether (1.2.1) satisfies the *relaxation property,* namely

$$\lim_{\tau \to \infty} \mathcal{F}(\phi^{t(\tau)}) = \mathcal{F}(\phi^t(0)^\dagger)$$

where $\phi^t(0)^\dagger$ is the constant history equal to $\phi^t(0)$. By the continuity of $\mathcal{F}$ this holds if and only if

$$\lim_{\tau \to \infty} \|\phi^{t(\tau)} - \phi^t(0)^\dagger\| = 0.$$

That is why the influence functions which meet (i)-(iii) are said to have the relaxation property. In addition we mention that, as proved in [19], [20], if $|\phi^t(s)|^p k(s)$ is bounded on $\mathbb{R}^+$, then influence functions $k(s)$ which are monotone decreasing for large s have the relaxation property.

It is possible to state the fading memory principle in an abstract way without introducing the influence function but simply specifying the properties of the Banach space associated with the fading memory norm (recollection); this general approach was developed by Coleman and Mizel [24]. Further aspects of fading memory concerning the chain rule are deferred to §1.3.

In Chapter 3 some results about viscoelasticity are derived by explicitly using the influence function. Indeed, in that context we need an additional, though reasonable, hypothesis on k. That is why we have presented the principle of fading memory through the influence function.

1.3. Fading memory space. As we shall see in the sequel, the description of continuous bodies involves functionals which depend on histories but with an explicit dependence on the present values. Accordingly, the concept of fading memory needs a more detailed framework which is specified through a fading memory space. This space constitutes the domain of the response functional (or functionals) in a way that will be made precise in Chapter 2.

Let k be an influence function of order $\alpha > 1$, which implies that k is integrable over $\mathbb{R}^+$. Let ${}_r\mathcal{H}$ be the Banach space

$$ {}_r\mathcal{H} = \left\{ {}_r\phi^t : \mathbb{R}^{++} \to W, \int_0^\infty |{}_r\phi^t(s)|^2 k(s) ds < \infty \right\}, $$

the norm $\|{}_r\phi^t\|_r$ of a past history ${}_r\phi^t$ defined by

$$ \|{}_r\phi^t\|_r^2 = \int_0^\infty |{}_r\phi^t(s)|^2 k(s) ds. $$

The Banach space ${}_r\mathcal{H}$ becomes a Hilbert space as soon as we introduce the inner product of two past histories ${}_r\phi_1^{t_1}$ and ${}_r\phi_2^{t_2}$ as

$$ ({}_r\phi_1^{t_1}, {}_r\phi_2^{t_2}) = \int_0^\infty {}_r\phi_1^{t_1}(s) \cdot {}_r\phi_2^{t_2}(s)\, k(s)\, ds. $$

The analogous inner product for histories, namely

$$ (\phi_1^{t_1}, \phi_2^{t_2}) = \phi_1(t_1) \cdot \phi_2(t_2) + \int_0^\infty \phi_1^{t_1}(s) \cdot \phi_2^{t_2}(s)\, k(s)\, ds, $$

ascribes to the set $\mathcal{H}$ of histories with a finite norm of the structure of Hilbert space, called *fading memory space* of (total) histories. Of course, the associated norm of a history ϕ^t is given by

$$ \|\phi^t\|^2 = |\phi(t)|^2 + \int_0^\infty |\phi^t(s)|^2 k(s) ds. \tag{1.3.1} $$

Unless suitable restrictions on ϕ are required, we regard the fading memory space $\mathcal{H}$ so defined as the domain of $\mathcal{F}$.

For later convenience we mention some properties of the fading memory space $\mathcal{H}$.

(I) If $\phi^t \in \mathcal{H}$ then the static continuation $\phi^{t(\tau)}$ is in $\mathcal{H}$.

The proof is immediate and involves essentially the monotone decreasing property of k for large s.

(II) If $\phi^t \in \mathcal{H}$ then the τ-section $\phi^t_{(\tau)}$ is in $\mathcal{H}$.

The proof involves the asymptotic property of k.

(III) The distance between the static continuation $\phi^{t(\tau)}$ of ϕ^t and the history $\phi^t(0)^\dagger$, with constant value $\phi^t(0)$, tends to zero in that

$$ \lim_{\tau\to\infty} \|\phi^{t(\tau)} - \phi^t(0)^\dagger\| = 0. \tag{1.3.2} $$

This follows from

$$ \|\phi^{t(\tau)} - \phi^t(0)^\dagger\| = \int_0^\infty |\phi^t(s) - \phi^t(0)|^2 k(s+\tau) ds $$

and the asymptotic property of k.

(IV) If $\phi^t \in \mathcal{H}$ then the retardation ϕ^t_γ is in $\mathcal{H}$.

The proof is based on the property $k(s) \leq k(\gamma s), \gamma \in (0,1)$, for large s.

Thus a *functional* maps $\mathcal{H}$ into a vector space Y (possibly Lin, $V, \mathbb{R}$). A functional f is said to be *linear* if

$$f(c_1\phi_1^{t_1} + c_2\phi_2^{t_2}) = c_1 f(\phi_1^{t_1}) + c_2 f(\phi_2^{t_2}), \qquad c_1, c_2 \in \mathbb{R},$$

for any pair of histories $\phi_1^{t_1}, \phi_2^{t_2} \in \mathcal{H}$, and *bounded* if there exists a positive constant M such that

$$|f(\phi^t)| \leq M\|\phi^t\|,$$

for any history $\phi^t \in \mathcal{H}$; here $|f|$ denotes the norm in Y. Let f be a bounded, linear (and then continuous) functional on the Hilbert space $\mathcal{H}$. Then, because of the Riesz theorem, there exists an element $g \in \mathcal{H}$ such that

$$f(\phi^t) = (g, \phi^t),$$

namely

$$f(\phi^t) = g(0)\phi^t(0) + \int_0^\infty g(s)\phi^t(s)k(s)ds, \tag{1.3.3}$$

for all histories ϕ^t in $\mathcal{H}$. Of course,

$$\|g\|^2 = |g(0)|^2 + \int_0^\infty |g(s)|^2 k(s)ds < \infty.$$

Let $\mathcal{G}(s) = g(s)k(s)$. Then, because $k(0) = 1$, we can write (1.3.3) as

$$f(\phi^t) = \mathcal{G}(0)\phi^t(0) + \int_0^\infty \mathcal{G}(s)\phi^t(s)ds \tag{1.3.4}$$

with the condition

$$\int_0^\infty |\mathcal{G}(s)|^2 k^{-1}(s)ds < \infty \tag{1.3.5}$$

ensuring the continuity of f.

Often histories are physically admissible only if suitable restrictions are satisfied; for example, $\phi^t(s) > 0, s \in \mathbb{R}^+$, if ϕ is the absolute temperature and $\det\phi^t(s) \neq 0, s \in \mathbb{R}^+$, if ϕ is the deformation gradient. In such a case the natural domain of the response functional $\mathcal{F}$ on the set of past histories is a nowhere dense cone C in S_k. It is assumed that the response functionals admit smooth extensions from the cone C to an open set in S_k in which elements of C form a dense set.

The principle of fading memory means that, for the materials under consideration, causes that occur in the distant past have less influence on the response than those occurring in the recent past. Operatively the principle of fading memory has been interpreted as a requirement of smoothness for the response functional $\mathcal{F}$. There is a thermodynamic theory (based on the Clausius–Duhem inequality; see Chapter 2) where the essential smoothness property for the response functional is the validity of the chain rule. Accordingly, following Mizel and Wang [97], we can state the fading memory assumption through the chain rule property (cf. also [38], §5.3.).

To make the formalism more evident we distinguish explicitly the dependence on the present value $\phi(t)$ from that on the past history ${}_r\phi^t$. Let $a, b \in W$ and $\lambda, \chi : \mathbb{R}^{++} \to W$. Moreover let the pairs (a, λ) and $(a + b, \lambda + \chi)$ be elements of $\mathcal{H}$. The functional $\mathcal{F}$ on $\mathcal{H}$ is said to be continuously differentiable if

$$\mathcal{F}(a+b, \lambda+\chi) = \mathcal{F}(a, \lambda) + D\mathcal{F}(a, \lambda)\, b + d\mathcal{F}(a, \lambda|\chi) + o(|b| + \|\chi\|) \tag{1.3.6}$$

where $D\mathcal{F}(a, \lambda) : W \to Y$ is a continuous functional; $d\mathcal{F}(a, \lambda|\chi)$ is the Fréchet differential, continuous in a, λ and linear in χ. Of course,

$$D\mathcal{F}(a, \lambda) = \frac{\partial \mathcal{F}}{\partial a}(a, \lambda).$$

A function ϕ on $\mathbb{R}$ is said to be *mild*[2] if it has two continuous derivatives $\dot{\phi}, \ddot{\phi}$ and, for every $t \in \mathbb{R}$, the past histories ${}_r\phi^t$, ${}_r\dot{\phi}^t$, ${}_r\ddot{\phi}^t$ are in S_k.[3] The following theorem provides the sought result.

THEOREM 1.3.1. *If ϕ is mild and the functional $\mathcal{F}(\phi(t), {}_r\phi^t)$ is continuously differentiable, then the function $F(t) = \mathcal{F}(\phi(t), {}_r\phi^t)$ is continuously differentiable and its derivative is given by*

$$\dot{F}(t) = D\mathcal{F}(\phi(t), {}_r\phi^t)\, \dot{\phi}(t) + d\mathcal{F}(\phi(t), {}_r\phi^t|{}_r\dot{\phi}^t). \tag{1.3.7}$$

The proof of this theorem is omitted because it is given, with slight differences, in [97] and [38].

Accordingly, the principle of fading memory can be stated by saying that the histories ϕ^t are mild and the response functional $\mathcal{F}$ is continuously differentiable.

1.4. Essentials of continuum mechanics. Following the standard scheme of continuum mechanics, we regard a body $\mathcal{B}$ as a continuous distribution

[2]In Day's notation mild means satisfying the chain rule conditions.

[3]It is worth observing that $\dot{\phi}^t(s) = \frac{d}{dt}\phi(t-s) = -\frac{d}{ds}\phi^t(s)$.

of material points. A body may occupy different regions of the Euclidean space $\mathcal{E}^3$ at different times. It is convenient to choose one of these regions as reference. We denote by $\mathcal{R} \subset \mathcal{E}^3$ such a reference region, usually called *reference placement* (or reference configuration) and label the material points by the position vector $\mathbf{X}$ they occupy in $\mathcal{R}$. To save writing, we usually identify $\mathbf{X}$ with the material point that occupies the position vector $\mathbf{X}$ in the reference placement $\mathcal{R}$.

The motion of $\mathcal{B}$ is given by a function $\mathbf{p} : \mathcal{R} \times \mathrm{I\!R} \to \mathcal{E}^3$ which assigns, for each $\mathbf{X} \in \mathcal{R}$, the position vector $\mathbf{x} \in \mathcal{E}^3$ at time t. Letting $\mathbf{p}(\mathcal{R})$ be the image of $\mathcal{R}$, the map $\mathbf{p}(\cdot, t) : \mathcal{R} \to \mathbf{p}(\mathcal{R})$, $t \in \mathrm{I\!R}$, is regarded as a diffeomorphism. We denote by $\mathbf{F} = \partial \mathbf{p}/\partial \mathbf{X}$[4] the *deformation gradient* and let $J = \det \mathbf{F}$; without any loss of generality we assume $J > 0$. The velocity $\mathbf{v}$ and the acceleration $\mathbf{a}$ are the first-order and second-order derivative of $\mathbf{p}$ with respect to t. The vector $\mathbf{u} = \mathbf{p}(\mathbf{X}) - \mathbf{X}$ is the *displacement* of $\mathbf{X}$.

To the deformation gradient $\mathbf{F}$ we can apply a general theorem of algebra.

POLAR DECOMPOSITION THEOREM. *Given any* $\mathbf{F} \in \mathrm{Lin}^+$, *there exist uniquely determined, positive definite, symmetric tensors* $\mathbf{U}, \mathbf{V}$ *and a rotation* $\mathbf{R}$ *such that*

$$\mathbf{F} = \mathbf{R}\,\mathbf{U} = \mathbf{V}\,\mathbf{R}.$$

The symmetric tensors $\mathbf{C} = \mathbf{F}^T\mathbf{F}$ and $\mathbf{B} = \mathbf{F}\,\mathbf{F}^T$ are related to $\mathbf{U}$ and $\mathbf{V}$ by

$$\mathbf{C} = \mathbf{U}^2, \qquad \mathbf{B} = \mathbf{V}^2;$$

$\mathbf{C}$ and $\mathbf{B}$ are called the *right* and *left Cauchy-Green tensors,* respectively. Sometimes it is useful to regard the present placement $\mathbf{p}(\mathcal{R})$ as the reference. For any $\tau \le t$ consider the previous placements $\mathcal{R}_\tau$ induced by the maps $\mathbf{p}(\cdot, \tau)$ and denote by $\mathbf{F}_t(\tau)$ the deformation gradient corresponding to the deformation $\mathbf{p}(\mathcal{R}) \to \mathcal{R}_\tau$; usually $\mathbf{F}_t(\tau)$ is called the *relative deformation gradient.* Through the chain rule for the differentiation of composite vector-valued functions, we obtain

$$\mathbf{F}_t(\tau) = \mathbf{F}(\tau)\,\mathbf{F}^{-1}(t).$$

Letting

$$\mathbf{C}_t(\tau) = \mathbf{F}_t^T(\tau)\,\mathbf{F}_t(\tau), \qquad \mathbf{B}_t(\tau) = \mathbf{F}_t(\tau)\,\mathbf{F}_t^T(\tau),$$

we call $\mathbf{C}_t$ and $\mathbf{B}_t$ the *relative right* and *left Cauchy-Green tensors.*

All pertinent fields on the body can be described by functions of $\mathbf{X}$ and t. Alternatively they can be described by functions of $\mathbf{x}$ and t. As t varies, we follow the evolution of the quantity under consideration at a material point in the first case; in the second one we follow the evolution at a point in the Euclidean space $\mathcal{E}^3$. The first case is referred to as material or Lagrangian

[4]In Cartesian components, $F_{iK} = \partial p_i/\partial X_K$.

description, the second one as spatial or Eulerian description. For any function $\phi(\mathbf{X}, t)$, the invertibility of the application

$$\mathbf{x} = \mathbf{p}(\mathbf{X}, t)$$

at any time t allows us to consider the corresponding Eulerian version, namely

$$\varphi(\mathbf{x}, t) := \phi(\mathbf{p}^{-1}(\mathbf{x}, t), t).$$

Hence we have

$$\frac{\partial \phi}{\partial \mathbf{X}} = \frac{\partial \phi}{\partial \mathbf{x}} \mathbf{F}.$$

As for the time derivative we have

$$\frac{\partial \phi}{\partial t} = \frac{\partial \varphi}{\partial t} + \mathbf{v} \cdot \nabla \varphi.$$

For formal simplicity we will not use different symbols for the two functions and, with a slight abuse of notation, write

$$\dot{\varphi} = \frac{\partial \varphi}{\partial t} + \mathbf{v} \cdot \nabla \varphi,$$

a superposed dot denoting the time derivative in the material description (material time derivative) and $\nabla = \partial / \partial \mathbf{x}$. The tensor field

$$\mathbf{L} = \nabla \mathbf{v}$$

is called the *velocity gradient.*[5] By applying the material gradient $\partial / \partial \mathbf{X}$ to $\mathbf{v}$ we obtain the useful identity

$$\dot{\mathbf{F}} = \mathbf{L}\,\mathbf{F}.$$

Any model of a body $\mathcal{B}$ must comply with general principles of continuum mechanics. Among them are the principles that mass and momentum are conserved, which means that appropriate balance laws must hold. Here we disregard bodies with internal structure (polar media, electromagnetic continua, mixtures, etc.) and then the balance laws can be given the following form.

A *mass density* field ρ is defined such that, for any motion $\mathbf{p}$, the mass $m(\Omega)$ of any subregion Ω of $\mathcal{R}$ is given by

$$m(\Omega) = \int_{\mathbf{p}(\Omega)} \rho \, dx.$$

[5]In Cartesian components, $L_{ij} = \partial v_i / \partial x_j$.

The balance (conservation) of mass is expressed by saying that $m(\Omega)$ is constant in time, viz.

$$\frac{d}{dt}m(\Omega) = 0$$

for any fixed $\Omega \in \mathcal{R}$. Then the observation that

$$\int_{\mathbf{p}(\Omega)} \phi\, dx = \int_{\Omega} \phi\, J\, dX \tag{1.4.1}$$

for any function ϕ on $\mathbf{p}(\mathcal{R}) \times \mathbb{R}$ or $\mathcal{R} \times \mathbb{R}$ shows that ρJ is a function of $\mathbf{X}$ only. We write

$$\rho J = \rho_0 \tag{1.4.2}$$

where $\rho_0 = \rho_0(\mathbf{X})$ takes the meaning of mass density in $\mathcal{R}$. Let $\nabla\cdot$ stand for the divergence operator, with respect to $\mathbf{x}$. Then time differentiation of (1.4.2) and the identity $\dot{J} = J\,\nabla \cdot \mathbf{v}$ yield

$$\dot{\rho} + \rho\nabla \cdot \mathbf{v} = 0. \tag{1.4.3}$$

Alternatively, (1.4.3) can be written as

$$\frac{\partial \rho}{\partial t} + \nabla \cdot (\rho \mathbf{v}) = 0. \tag{1.4.4}$$

So we have three local forms of balance of mass. Equation (1.4.2) provides directly the mass density in the current configuration, ρ in terms of the reference mass density ρ_0 and the deformation through J, while (1.4.3) and (1.4.4) are the differential forms in the Lagrangian and Eulerian description, respectively.

Incidentally, for any C^1 field ϕ and region $\Omega \subset \mathcal{R}$ we have the result (*Reynold's transport theorem*)

$$\frac{d}{dt}\int_{\mathbf{p}(\Omega)} \phi\, dx = \int_{\mathbf{p}(\Omega)} (\dot{\phi} + \phi\nabla \cdot \mathbf{v})dx.$$

The proof follows at once by applying (1.4.1) twice. As a corollary we obtain that, for any C^1 field Φ and region $\Omega \subset \mathcal{R}$,

$$\frac{d}{dt}\int_{\mathbf{p}(\Omega)} \rho\,\Phi\, dx = \int_{\mathbf{p}(\Omega)} \rho\,\dot{\Phi}\, dx. \tag{1.4.5}$$

The force on a region $\Omega \subset \mathcal{R}$ is taken to consist of contact forces, by the remaining region $\mathcal{R} \setminus \Omega$ (and possibly the environment of the body), and body forces by the environment of the body. Following Cauchy's hypothesis, we assume that the contact forces are expressed by a surface force density $\mathbf{s}(\mathbf{n}, \mathbf{x}, t)$, $\mathbf{n}$ being the unit normal outgoing from $\partial\Omega$, $\mathbf{x}$ the position vector on $\partial\Omega$. Then for any region $\Omega \subset \mathcal{R}$ we write the balance of linear momentum as

$$\frac{d}{dt}\int_{\mathbf{p}(\Omega)} \rho\,\mathbf{v}\,dx = \int_{\partial\mathbf{p}(\Omega)} \mathbf{s}(\mathbf{n})da + \int_{\mathbf{p}(\Omega)} \rho\,\mathbf{b}\,dx.$$

Then Cauchy's theorem proves the existence of the *Cauchy stress tensor* $\mathbf{T}(\mathbf{x}, t)$ such that

$$\mathbf{s}(\mathbf{n}) = \mathbf{T}\,\mathbf{n}.$$

The application of (1.4.5) and the arbitrariness of Ω implies that the balance of linear momentum is expressed (in local form) by the *equation of motion*

$$\rho\dot{\mathbf{v}} = \nabla \cdot \mathbf{T} + \rho\,\mathbf{b}, \tag{1.4.6}$$

the divergence of a second-order tensor being relative to the second index.

An analogous integral balance law for angular momentum and use of (1.4.6) lead to the conclusion that $\mathbf{T}$ is a symmetric tensor, namely

$$\mathbf{T} = \mathbf{T}^T.$$

The previous scheme is sufficient to describe purely mechanical phenomena. Additional quantities and balance laws are necessary to account also for thermal aspects. Let $Q_\Omega(t)$ be the heat supplied to the region occupied by Ω at time t, i.e., $\mathbf{p}(\Omega)$. Such heat can be expressed as

$$Q_\Omega = -\int_{\partial\mathbf{p}(\Omega)} q\,da + \int_{\mathbf{p}(\Omega)} \rho\,r\,dx$$

where r is the heat supply, per unit mass, from the external world $\mathcal{E}^3 \setminus \Omega$. Let h be the rate, per unit mass, at which heat is absorbed; it is

$$Q_\Omega = \int_{\mathbf{p}(\Omega)} \rho\,h\,dx.$$

By analogy with Cauchy's theorem we can show that

$$q = \mathbf{q} \cdot \mathbf{n},$$

the vector field $\mathbf{q}$ being called the *heat flux* (vector). Then, by the divergence theorem and the arbitrariness of Ω we have

$$\rho\,h = -\nabla \cdot \mathbf{q} + \rho\,r.$$

To establish a useful connection with other, more traditional procedures we examine also the balance of energy (first law of thermodynamics) and the balance of entropy (second law of thermodynamics). We denote by ϵ the *internal energy* density so that $\frac{1}{2}v^2 + \epsilon$ is the total energy per unit mass. In addition to the power of the body force $\rho\,\mathbf{b}\cdot\mathbf{v}\,dx$ and the contact force $\mathbf{s}\cdot\mathbf{v}da$, we have to account for an exchange of energy, of nonmechanical character, between the region under consideration Ω and the environment $\mathcal{E}^3 \setminus \Omega$ of Ω. Owing to (1.4.6) we have

$$\rho\, h = \rho\dot{\epsilon} - \mathbf{T}\cdot\mathbf{L},$$

$$\rho\dot{\epsilon} = \mathbf{T}\cdot\mathbf{L} - \nabla\cdot\mathbf{q} + \rho\, r. \tag{1.4.7}$$

By the symmetry of $\mathbf{T}$ we have

$$\mathbf{T}\cdot\mathbf{L} = \mathbf{T}\cdot\mathbf{D};$$

$\mathbf{D} = \operatorname{sym}\mathbf{L}$ is often called *stretching* or *rate-of-strain tensor.* Let η be the *entropy* density and θ the *absolute temperature.* The second law of thermodynamics is often regarded as expressed by the (local) *Clausius-Duhem inequality*

$$\rho\dot{\eta} + \nabla\cdot\left(\frac{1}{\theta}\mathbf{q}\right) - \frac{\rho\, r}{\theta} \geq 0. \tag{1.4.8}$$

In terms of the *free energy density* $\psi = \epsilon - \theta\,\eta$, upon substitution for r from (1.4.7) we can write (1.4.8) as

$$-\rho(\dot{\psi} + \eta\,\dot{\theta}) + \mathbf{T}\cdot\mathbf{D} - \frac{1}{\theta}\mathbf{q}\cdot\mathbf{g} \geq 0, \tag{1.4.9}$$

where $\mathbf{g} = \nabla\theta$. It is by now customary to follow Coleman and Noll's viewpoint [27], [28] and interpret (1.4.9), or alternative formulations of the second law of thermodynamics, as a condition which must hold for every thermodynamic process. The operative, mathematical meaning of this statement is made clear in the next chapter. Conceptually, the main feature of the formulation (1.4.8) or (1.4.9) of the second law is that the existence of the entropy function is assumed at the outset. A weaker, integral formulation of the second law, free from this assumption, involves cyclic processes between equilibrium states: in any cyclic process between times t_1 and t_2, which starts from equilibrium at a particle $\mathbf{X}$, the inequality

$$\int_{t_1}^{t_2}\left[-\frac{1}{\rho}\nabla\cdot\left(\frac{1}{\theta}\mathbf{q}\right) + \frac{r}{\theta}\right]dt \leq 0 \tag{1.4.10}$$

holds with the argument evaluated at the particle $\mathbf{X}$. The inequality (1.4.10) is usually called *Clausius inequality.* The definitions of the (cyclic) process and the (equilibrium) state are provided in the next chapter.

Chapter 2

Thermodynamics of Simple Materials

2.1. Thermodynamic processes and states. At the bottom, any thermodynamic theory is based on the notions of state and process. Here we provide these notions by following Noll's theory [106] and the subsequent elaborations by Coleman and Owen [29], Fabrizio [45], and Fabrizio and Giorgi [48].

For any material point $\mathbf{X}$ in a body $\mathcal{B}$ and time t in a suitable interval in $\mathbb{R}$, we define the *configuration* as the pair

$$C(\mathbf{X},t) = (\mathbf{F}(\mathbf{X},t), \theta(\mathbf{X},t)) \tag{2.1.1}$$

of the deformation gradient $\mathbf{F}$ and the absolute temperature θ. Whenever no ambiguity can arise, the dependence on $\mathbf{X}$ is understood and not written.

Let E be an open, connected subset of $\mathrm{Lin}\times\mathbb{R}\times V$. A map $P:[0,d_P)\to E$, piecewise continuous[6] on $[0,d_P)$, is called *process* of duration $d_P\in\mathbb{R}^+$. Each restriction of P to an interval $[t_1,t_2)\subset[0,d_P)$ is called *segment* of P. In particular, we denote by P_t the restriction of P to the interval $[0,t)$, $t<d_P$. Letting P_1 and P_2 be two processes we define the process $P_1 * P_2$ as

$$P_1 * P_2 = \begin{cases} P_1(t), & t\in[0,d_{P_1}), \\ P_2(t-d_{P_1}), & t\in[d_{P_1}, d_{P_1}+d_{P_2}); \end{cases}$$

the process $P_1 * P_2$ is called *composition* of P_2 with P_1. Here we consider *thermokinetic* processes defined, for every $t\in[0,d_P)$, as

$$P(t) = (\mathbf{L}(t), \dot\theta(t), \mathbf{g}(t)). \tag{2.1.2}$$

[6]A function f is said piecewise continuous on $[0,d)$ if $\lim f(\xi)=f(t)$ as $\xi\to t^+$ and $\lim f(\xi)$ exists as $\xi\to t^-$ for each $t\in[0,d)$ and $\lim f(\xi)=f(t)$ as $\xi\to t^-$ at all but a finite number of points in $[0,d)$.

The symbol $\mathcal{C}$ denotes the set of all configurations which are accessible to the body $\mathcal{B}$ at the point $\mathbf{X}$ under consideration. The set Π of all accessible processes is such that if $P \in \Pi$ then every segment of P is in Π. Moreover, if $P_1, P_2 \in \Pi$ then $P_1 * P_2 \in \Pi$.

Any process $P \in \Pi$ can be viewed as the *input* of $\mathcal{B}$. The *output* U of $\mathcal{B}$ is an element of Sym $\times$ $\mathbb{R} \times V$ and is defined as

$$U = (\mathbf{T}, h, \mathbf{q}). \tag{2.1.3}$$

DEFINITION 2.1.1. A *simple material element,* at any point $\mathbf{X} \in \mathcal{B}$, is the ordered array $(\mathcal{C}, \Pi, \Sigma, \hat{C}, \varrho, \hat{U})$; the entries of the array are specified as follows.

(1) $\mathcal{C}$ is an open, connected subset of $\text{Lin}^+ \times \mathbb{R}^+$.

(2) Π is the set of piecewise-continuous thermokinetic processes.

(3) Σ is a metric space which is called the state space; its elements, the states, are denoted by σ.

(4) $\hat{C} : \Sigma \to \mathcal{C}$ is a projection so that we can write $\Sigma = \mathcal{C} \times \Sigma'$; incidentally this means that any state σ comprises the configuration C.

(5) $\varrho : \Sigma \Diamond \Pi \to \Sigma$ is called *evolution function* (or state-transformation function); its domain $\Sigma \Diamond \Pi$ is a subset of $\Sigma \times \Pi$ such that any element is the pair (σ, P) of a state σ and a process P which can occur by starting from σ. The function ϱ is such that if $(\sigma, P_1) \in \Sigma \Diamond \Pi$ and $(\varrho(\sigma, P_1), P_2) \in \Sigma \Diamond \Pi$ then $(\sigma, P_1 * P_2) \in \Sigma \Diamond \Pi$ and $\varrho(\sigma, P_1 * P_2) = \varrho(\varrho(\sigma, P_1), P_2)$.

(6) $\hat{U} : \Sigma \times E \to \text{Sym} \times \mathbb{R} \times V$ maps the pair $(\sigma(t), P(t))$ of the state σ and the process P at a time t into the output U at the time t. Specifically, $\hat{U}$ consists of the triplet $\hat{\mathbf{T}}, \hat{h}, \hat{\mathbf{q}}$ such that

$$\hat{\mathbf{T}} : \Sigma \times E \to \text{Sym}, \qquad \hat{h} : \Sigma \times E \to \mathbb{R}, \qquad \hat{\mathbf{q}} : \Sigma \times E \to V. \tag{2.1.4}$$

In other words, $U(t)$ depends on the state $\sigma(t)$ and the present value of the input, $P(t)$.

As a first example of simple material element, consider a thermoelastic material whose output U is characterized by (the constitutive equations)

$$\mathbf{T} = \mathbf{T}(\mathbf{F}, \theta), \qquad h = a(\mathbf{F}, \theta)\dot{\theta} + \mathbf{B}(\mathbf{F}, \theta) \cdot \mathbf{L}, \qquad \mathbf{q} = -\mathbf{K}(\mathbf{F}, \theta)\mathbf{g} \tag{2.1.5}$$

where $\mathbf{B}, \mathbf{K} \in \text{Lin}$ and $\mathbf{K}$ is positive definite. The state coincides with the configuration C while the process has the usual form (2.1.2). To avoid any ambiguity we write

$$P(t) = (\mathbf{L}_P(t), \dot{\theta}_P(t), \mathbf{g}_P(t)) \tag{2.1.6}$$

where the subscript P is a reminder that the time dependence for $\mathbf{L}, \dot{\theta}$, and $\mathbf{g}$ is that induced by the process P. It is a trivial matter to determine the evolution function ϱ. Let $\sigma^i = (\mathbf{F}_0, \theta_0)$ be the initial state at $t = 0$. Then the current state $\sigma(t) = (\mathbf{F}(t), \theta(t)) := \varrho(\sigma^i, P_t)$ is the solution to the Cauchy problem

$$\begin{cases} \frac{d}{dt}\mathbf{F}(t) = \mathbf{L}_P(t)\mathbf{F}(t), & \mathbf{F}(0) = \mathbf{F}_0, \\ \frac{d}{dt}\theta(t) = \dot{\theta}_P(t), & \theta(0) = \theta_0, \end{cases}$$

with $t \in [0, d_P)$, where $\mathbf{L}$ and $\dot{\theta}$, as elements of P, are given functions of t in $[0, d_P)$. The output U follows immediately from (2.1.5).

Incidentally, it is apparent from (2.1.5) that $\mathbf{T}$ is a state function, in that it depends on the state σ only, while h and $\mathbf{q}$ depend also on the process P; this dependence is linear for h and $\mathbf{q}$.

Another example is provided by the Newtonian viscous fluid whose output function $\hat{U}$ is given by

(2.1.7)
$$\mathbf{T} = -p(\rho, \theta)\mathbf{1} + 2\mu\mathbf{D} + \lambda(\mathrm{tr}\mathbf{D})\mathbf{1}, \quad h = c(\mathbf{F}, \theta)\dot{\theta} + \mathbf{G}(\mathbf{F}, \theta)\cdot\mathbf{D}, \quad \mathbf{q} = \mathbf{K}(\mathbf{F}, \theta)\mathbf{g}.$$

Here μ is the shear viscosity, $\lambda + 2\mu/3$ the bulk viscosity. They both can depend on θ and $\mathbf{F}$ through ρ. Again the state σ coincides with the configuration $(\mathbf{F}, \theta)$. The evolution function ϱ is determined as in the previous example.

As a further example, consider a material with fading memory whose output function is characterized by the constitutive equations

(2.1.8)
$$\mathbf{T} = \hat{\mathbf{T}}(\mathbf{F}^t, \theta^t), \qquad h = \hat{a}(\mathbf{F}^t, \theta^t)\dot{\theta} + \hat{\mathbf{B}}(\mathbf{F}^t, \theta^t) \cdot \mathbf{L}, \qquad \mathbf{q} = \hat{\mathbf{q}}(\mathbf{F}^t, \theta^t, \mathbf{g}).$$

Here the state σ is the pair $(\mathbf{F}^t, \theta^t)$ of the histories $\mathbf{F}^t, \theta^t$ which are elements of the fading memory space

$$H = \left\{\mathbf{F}^t, \theta^t : \mathbb{R}^+ \to \mathrm{Lin}^+ \times \mathbb{R}^+; \int_0^\infty [|\mathbf{F}^t(s)|^2 + |\theta^t(s)|^2]\, k(s)\, ds < \infty\right\}$$

where k is the usual influence function. It is a routine matter to determine $\hat{C}$ and ϱ.

The final example is provided by materials with internal variables [22]. To be specific, letting $\boldsymbol{\xi} \in \mathbb{R}^n$ be the set of internal variables, we write the constitutive equations as

$$\mathbf{T} = \mathbf{T}(\mathbf{F}, \theta, \boldsymbol{\xi}), \qquad h = a(\mathbf{F}, \theta, \boldsymbol{\xi})\dot{\theta} + \mathbf{B}(\mathbf{F}, \theta, \boldsymbol{\xi}) \cdot \mathbf{L}, \qquad \mathbf{q} = \mathbf{q}(\mathbf{F}, \theta, \mathbf{g}, \boldsymbol{\xi})$$

along with the growth equation for $\boldsymbol{\xi}$,

$$\dot{\boldsymbol{\xi}} = \mathbf{f}(\mathbf{F}, \theta, \mathbf{g}, \boldsymbol{\xi}).$$

Here the state σ is the triplet $(\mathbf{F}, \theta, \boldsymbol{\xi})$ and the process is $P(t) = (\mathbf{L}_P(t), \dot{\theta}_P(t), \mathbf{g}_P(t))$. The evolution function ϱ determines the current state $\sigma(t)$ in terms of the initial state σ^i, the segment P_t, and the (given) growth function $\mathbf{f}$.

2.2. First and second law of thermodynamics. Given a process $P \in \Pi$, let

$$\mathcal{D}(P) := \{\sigma \in \Sigma; (\sigma, P) \in \Sigma \diamondsuit \Pi\}.$$

A pair (σ, P), $\sigma \in \mathcal{D}(P)$, is called a *cycle* if

$$\varrho(\sigma, P) = \sigma.$$

In other words, the final state produced by the process P coincides with the initial state. In such a case we say also that the process P is cyclic.

The mass density ρ is determined by the state σ; indeed, the dependence on σ is through $\det\mathbf{F}$ only. To save writing, from now on we write $\rho(t)$ as an abbreviation for $\rho(\sigma(t))$.

First law. *For every cycle* $(\sigma, P) \in \Sigma \diamondsuit \Pi$ *the equality*

$$\int_0^{d_P} \left[\hat{h}(\sigma(t), P(t)) + \frac{1}{\rho(t)} \hat{\mathbf{T}}(\sigma(t), P(t)) \cdot \mathbf{L}(t)\right] dt = 0$$

holds with $\sigma(t) = \varrho(\sigma, P_t)$.

Let $e : \Sigma \diamondsuit \Pi \to \mathbb{R}$ be defined by

$$e(\sigma, P) = \int_0^{d_P} \left[\hat{h}(\sigma(t), P(t)) + \frac{1}{\rho(t)} \hat{\mathbf{T}}(\sigma(t), P(t)) \cdot \mathbf{L}(t)\right] dt. \tag{2.2.1}$$

Given $\sigma_0, \sigma \in \Sigma$ it is convenient to consider the set [29]

$$\mathcal{E}(\sigma_0, \sigma) := \{e(\sigma_0, P), P \in \Pi : \varrho(\sigma_0, P) = \sigma\}.$$

LEMMA 2.2.1. *A necessary condition for the validity of the first law is that, for any* $\sigma_0, \sigma \in \Sigma$, *the set* $\mathcal{E}(\sigma_0, \sigma)$ *is bounded and consists of the only value* $e(\sigma_0, \bar{P})$, $\bar{P} \in \Pi$ *being any process such that* $\varrho(\sigma_0, \bar{P}) = \sigma$.

Proof. By the first law, for any cyclic process $P \in \Pi$ such that $\varrho(\sigma_0, P) = \sigma_0$, we have

$$e(\sigma_0, P) = 0.$$

Given $\sigma \in \Sigma$, let $\bar{P} \in \Pi$ be such that $\sigma \in \mathcal{D}(\bar{P})$ and $\varrho(\sigma, \bar{P}) = \sigma_0$. Then, by the additivity of the function e, for any process $P \in \Pi$ such that $\varrho(\sigma_0, P) = \sigma$ the function e satisfies

$$e(\sigma_0, P) + e(\sigma, \bar{P}) = 0. \tag{2.2.2}$$

Hence $e(\sigma_0, P)$ is always equal to $-e(\sigma, \bar{P})$ for any $P : \varrho(\sigma_0, P) = \sigma$. It is a trivial matter to show that the conclusion is independent of the choice of the particular process $\bar{P}$ in that if $\bar{\bar{P}}$ is such that $\varrho(\sigma_0, \bar{\bar{P}}) = \sigma$, then $e(\sigma_0, \bar{\bar{P}}) = e(\sigma_0, \bar{P})$. It is thus proved that $\mathcal{E}$ consists of only one value. □

DEFINITION 2.2.1. A function $A : \Sigma \to \mathbb{R}$ is said to be a *potential* for $e(\sigma, P)$ if

$$A(\sigma_2) - A(\sigma_1) = e(\sigma_1, P)$$

for every pair of states $\sigma_1, \sigma_2 \in \Sigma$ and process $P \in \Pi$ such that $\varrho(\sigma_1, P) = \sigma_2$.

THEOREM 2.2.1. *A necessary condition for the validity of the first law is that there exists a potential* $\epsilon : \Sigma \to \mathbb{R}$ *for* $e(\sigma, P)$.

Proof. By Definition 2.2.1 of $e(\sigma, P)$, ϵ being a potential for $e(\sigma, P)$ means that for any pair of states $\sigma_1, \sigma_2 \in \Sigma$ and any process $P \in \Pi$, with $\varrho(\sigma_1, P) = \sigma_2$,

$$\epsilon(\sigma_2) - \epsilon(\sigma_1) = \int_0^{d_P} \left[\hat{h}(\sigma(t), P(t)) + \frac{1}{\rho(t)} \hat{\mathbf{T}}(\sigma(t), P(t)) \cdot \mathbf{L}(t) \right] dt$$

where $\sigma(t) = \varrho(\sigma_1, P_t)$.

Given $\sigma_0 \in \Sigma$ and $P \in \Pi$ such that $\sigma = \varrho(\sigma_0, P)$, define the function

$$\epsilon(\sigma) = e(\sigma_0, P). \tag{2.2.3}$$

Now we prove that the function $\epsilon(\sigma)$ is a potential for e. For any two states $\sigma_1, \sigma_2 \in \Sigma$ we have

$$\epsilon(\sigma_2) - \epsilon(\sigma_1) = e(\sigma_0, P_2) - e(\sigma_0, P_1)$$

whenever $P_2, P_1 \in \Pi$ satisfy $\varrho(\sigma_0, P_2) = \sigma_2, \varrho(\sigma_0, P_1) = \sigma_1$. By the additivity of the function e, letting P be any process such that $\varrho(\sigma_1, P) = \sigma_2$ we have

$$\epsilon(\sigma_2) - \epsilon(\sigma_1) = e(\sigma_0, P_1 * P) - e(\sigma_0, P_1) = e(\sigma_1, P). \tag{2.2.4}$$

The observation that (2.2.4) holds for arbitrary states $\sigma_1, \sigma_2 \in \Sigma$ and process $P \in \Pi$ such that $\varrho(\sigma_1, P) = \sigma_2$ completes the proof. □

On the basis of (2.2.1) and (2.2.3) we regard the function $\epsilon(\sigma)$ as the internal energy (density). Of course the internal energy ϵ is determined to within an additive constant. This arbitrariness can be removed by regarding the internal energy at a state σ_0 as the reference value.

It follows from Theorem 2.2.1 that, at any time t where P is continuous,

$$\dot{\epsilon}(\sigma(t)) = \hat{h}(\sigma(t), P(t)) + \frac{1}{\rho(t)} \hat{\mathbf{T}}(\sigma(t), P(t)) \cdot \mathbf{L}(t).$$

By the definition of h, viz.

(2.2.5)
$$\rho h = -\nabla \cdot \mathbf{q} + \rho r,$$

we have the relation

(2.2.6)
$$\rho \dot{\epsilon} = -\nabla \cdot \mathbf{q} + \mathbf{T} \cdot \mathbf{L} + \rho r$$

which identifies ϵ with the internal energy of standard continuum thermomechanics (cf. (1.3.7)).

The content of the second law is given the following form.

Second law. *For every cycle $(\sigma, P) \in \Sigma \diamondsuit \Pi$ the inequality*

$$\int_0^{d_P} \left[\frac{\hat{h}(\sigma(t), P(t))}{\theta(t)} + \frac{1}{\rho(t)\theta^2(t)} \hat{\mathbf{q}}(\sigma(t), P(t)) \cdot \mathbf{g}(t) \right] dt \leq 0$$

holds with $\sigma(t) = \varrho(\sigma, P_t)$.

To derive consequences of the second law it is convenient to consider the function $s : \Sigma \diamondsuit \mathbb{R} \to \mathbb{R}$, defined by

$$s(\sigma, P) = \int_0^{d_P} \left[\frac{\hat{h}(\sigma(t), P(t))}{\theta(t)} + \frac{1}{\rho(t)\theta^2(t)} \hat{\mathbf{q}}(\sigma(t), P(t)) \cdot \mathbf{g}(t) \right] dt,$$

and the set

$$\mathcal{S}(\sigma_0, \sigma) = \{s(\sigma_0, P), P \in \Pi : \varrho(\sigma_0, P) = \sigma\}.$$

Moreover it is convenient to introduce the notion of upper potential [29].

DEFINITION 2.2.2. A function $B : \Sigma \to \mathbb{R}$ is said to be an upper potential for $s(\sigma, P)$ if

$$B(\sigma_2) - B(\sigma_1) \geq s(\sigma_1, P)$$

for every pair of states $\sigma_1, \sigma_2 \in \Sigma$ and process $P \in \Pi$ such that $\varrho(\sigma_1, P) = \sigma_2$.

THEOREM 2.2.2. *A necessary condition for the validity of the second law is that there exists at least one upper potential for $s(\sigma, P)$.*

Proof. Let $\sigma_0, \sigma \in \Sigma$ and $\bar{P} \in \Pi$ with $\varrho(\sigma, \bar{P}) = \sigma_0$. For every $P \in \Pi$ with $\varrho(\sigma_0, P) = \sigma$ it follows from the second law that

$$s(\sigma_0, P * \bar{P}) \leq 0.$$

Hence

$$s(\sigma_0, P) \leq -s(\sigma_0, \bar{P})$$

for every $P \in \Pi$ with $\varrho(\sigma_0, P) = \sigma$. Then there exists $M \in \mathbb{R}$ such that

$$s(\sigma_0, P) \leq M$$

whereby the set $\mathcal{S}(\sigma_0, \sigma)$ is bounded.

Given $\sigma_0 \in \Sigma$ let $\eta^s(\sigma) = \sup\{s(\sigma_0, P),\, P \in \Pi : \varrho(\sigma_0, P) = \sigma\}$. Then there exists $\varepsilon > 0$ and $P_\varepsilon \in \Pi$ with $\varrho(\sigma_0, P_\varepsilon) = \sigma_1$ such that

$$s(\sigma_0, P_\varepsilon) > \eta^s(\sigma_1) - \varepsilon.$$

Consider $\sigma_2 \in \Pi$ and $P \in \Pi$ with $\varrho(\sigma_1, P) = \sigma_2$. Then

$$\eta^s(\sigma_2) \geq s(\sigma_0, P_\varepsilon * P) = s(\sigma_0, P_\varepsilon) + s(\sigma_1, P)$$

whence

$$\eta^s(\sigma_2) - \eta^s(\sigma_1) > s(\sigma_1, P) - \varepsilon.$$

By the arbitrariness of ε, we have the relation

$$\eta^s(\sigma_2) - \eta^s(\sigma_1) \geq s(\sigma_1, P)$$

which proves the upper potential property of $\eta^s(\sigma)$ for $s(\sigma, P)$. □

The upper potential $\eta^s(\sigma)$ is called entropy.

It is worth remarking that in general the upper potential $\eta^s(\sigma)$ is not uniquely determined. We can fix the value of $\eta^s(\sigma)$ at a reference state σ_0 and denote the upper potential by $\eta(\sigma)$. Yet even $\eta(\sigma)$ is not uniquely determined and we have

$$\eta^s(\sigma) \geq \eta(\sigma).$$

For completeness and useful reference, we give the explicit statement of the second law when the processes under consideration can be regarded as isothermal. Observe that, by (2.2.5) and (2.2.6),

$$\rho h = \rho \dot{\epsilon} - \mathbf{T} \cdot \mathbf{L}.$$

Substitution yields

$$s(\sigma, P) = \int_0^{d_P} \left[\frac{\dot{\hat{\epsilon}}(\sigma(t), P(t))}{\theta(t)} - \frac{1}{\rho(t)} \hat{\mathbf{T}}(\sigma(t), P(t)) \cdot \mathbf{L}(t) \right.$$
$$\left. + \frac{1}{\rho(t)\theta^2(t)} \hat{\mathbf{q}}(\sigma(t), P(t)) \cdot \mathbf{g}(t) \right] dt.$$

If the process P is cyclic and θ is constant and uniform, i.e., $\mathbf{g} = 0$, then the only significant contribution is given by $\mathbf{T} \cdot \mathbf{L}$. Hence we have the following statement.

Second law for isothermal processes. *For every cycle* $(\sigma, P) \in \Sigma \Diamond \Pi$ *the inequality*

$$\int_0^{d_P} \frac{1}{\rho(t)} \hat{\mathbf{T}}(\sigma(t), P(t)) \cdot \mathbf{L}(t) \geq 0$$

holds with $\sigma(t) = \varrho(\sigma, P_t)$ *and* $P(t) = (\mathbf{L}(t), 0, \mathbf{0})$.

2.3. Second law for approximate cycles. For materials with fading memory, cycles are quite rare because usually the material gets to a state which, though close to the initial one, is different from it. Then, to make the thermodynamic theory widely applicable, a broader class of processes are considered as the counterpart of cyclic processes. Such is the case of approximately cyclic processes and of approximate cycles, characterized as follows.

Let $\mathcal{O}_\nu(\sigma)$ stand for the ν-neighbourhood of σ consisting of the set of elements $\bar{\sigma}$ such that $\|\bar{\sigma} - \sigma\| < \nu$. A pair $(\sigma, P) \in \Sigma \times \Pi$ is called a ν-approximate cycle if $\varrho(\sigma, P) \in \mathcal{O}_\nu(\sigma)$.

Letting $\sigma(t) = \varrho(\sigma, P_t)$ we state the extension of the second law for approximate cycles as follows [29].

Second law for approximate cycles. *For every* $\varepsilon > 0$ *there exists* $\nu_\varepsilon > 0$ *such that*

$$\int_0^{d_P} \left[\frac{\hat{h}(\sigma(t), P(t))}{\theta(t)} + \frac{1}{\rho(t)\,\theta^2(t)} \hat{\mathbf{q}}(\sigma(t), P(t)) \cdot \mathbf{g}(t) \right] dt < \varepsilon$$

for any state σ *and process* P *with* $\varrho(\sigma, P) \in \mathcal{O}_{\nu_\varepsilon}(\sigma)$.

For later convenience it is worth introducing the dissipation.

DEFINITION 2.3.1. The *dissipation* functional $\hat{\gamma} : \Sigma \times E \to \mathbb{R}$ is given by

$$\hat{\gamma}(\sigma(t), P(t)) = \frac{1}{\theta(t)} \hat{h}(\sigma(t), P(t)) + \frac{1}{\rho(t)\theta^2(t)} \hat{\mathbf{q}}(\sigma(t), P(t)) \cdot \mathbf{g}(t).$$

By (2.2.5) and (2.2.6) the value of the dissipation, γ, is given by

$$\gamma = \frac{1}{\theta}\dot{\epsilon} - \frac{1}{\rho\theta}\mathbf{T} \cdot \mathbf{L} + \frac{1}{\rho\theta^2}\mathbf{q} \cdot \mathbf{g}. \tag{2.3.1}$$

The statement of the second law simplifies when, as is the case in linear viscoelasticity, isothermal processes are involved. First, the contribution to the dissipation due to heat conduction vanishes. Second, the integral of $\dot{\epsilon}/\theta$ in $[0, d)$ gives $[\hat{\epsilon}(\sigma(d)) - \hat{\epsilon}(\sigma(0))]/\theta$. Provided only that the functional $\hat{\epsilon}$ is continuous, we can have $|\hat{\epsilon}(\sigma(d)) - \hat{\epsilon}(\sigma(0))|$ as small as we please by considering a small enough neighbourhood $\mathcal{O}_\nu(\sigma)$ of the initial state σ. Then in the case of isothermal processes the second law for approximate cycles becomes the following.

Work inequality. *For every* $\varepsilon > 0$ *there exists* $\nu_\varepsilon > 0$ *such that*

$$\int_0^d \frac{1}{\rho(t)} \hat{\mathbf{T}}(\sigma(t), P(t)) \cdot \mathbf{L}(t)\, dt > -\varepsilon$$

for any state σ *and process* P *with* $\varrho(\sigma, P) \in \mathcal{O}_{\nu_\varepsilon}(\sigma)$.

There are cases where the set of cycles is quite large, relative to the set of approximate cycles, in the sense that no additional information arises from the second law for approximate cycles, beyond that from the second law for cycles. As we shall see in the next chapter, such is the case for the linear theory of viscoelasticity.

2.4. Reversible processes. Traditional books on thermodynamics distinguish between reversible and irreversible processes for homogeneous bodies. This distinction has not been given much attention, if any, in continuum thermomechanics. Standard statements of the second law do not involve the notion of reversibility in an operative way. Yet in our minds the characterization of reversible, and then irreversible, processes is of fundamental importance when dealing with the second law of thermodynamics. In a paper of ours [101] we have shown how a standard statement of the second law leads to a thermodynamic restriction which allows for the nonuniqueness of the solution to the quasi-static problem (cf. §4.3.). That is why we devote this section to the pertinent definitions and properties of reversible processes [55].

DEFINITION 2.4.1. For any process P on $[0, d)$ the reverse process $\tilde{P}$, on $[0, d)$, is defined as

(2.4.1)
$$\tilde{P}(t) = (\tilde{\mathbf{L}}(t), \dot{\tilde{\theta}}(t), \tilde{\mathbf{g}}(t)) = (-\mathbf{L}(d-t), -\dot{\theta}(d-t), -\mathbf{g}(d-t)), \qquad \forall t \in [0, d).$$

An alternative definition of reverse process might be given with $\tilde{\mathbf{g}}(t) = \mathbf{g}(d-t)$. This possibility is commented on in Remark 2.4.1.

DEFINITION 2.4.2. A pair (σ, P) is said to be reversible if the reverse process $\tilde{P}$ satisfies

(a)
$$\varrho(\sigma, P_t) = \varrho(\varrho(\sigma, P), \tilde{P}_{d-t}), \qquad \forall t \in [0, d_P),$$

(b)
$$\hat{\gamma}(\sigma(t), P(t)) = -\hat{\gamma}(\sigma(t), \tilde{P}(d-t)), \qquad \forall t \in [0, d_P).$$

A process P which does not satisfy (a) and (b) is said to be irreversible.

Borrowing from the physical terminology, we can phrase this definition by saying that a process is reversible if the state is even under time reversal while the dissipation is odd.

For ease in writing, the state $\varrho(\varrho(\sigma, P), \tilde{P}_{d-t})$ will be denoted by $\tilde{\sigma}(d-t)$.

It is worth observing that the properties (a) and (b) involve the state and the dissipation at any time t in $[0, d)$. Accordingly, Definition 2.4.1 is in fact the characterization of locally reversible processes; to save writing the specification "locally" will be omitted throughout.

Incidentally, often the term "reversible process" is used for processes that are performed in such a way that eventually both the system and the local surroundings may be restored to their initial states, without producing any changes in the rest of the universe (cf. [132], Chap. 8). The properties (a) and (b) provide an operative definition of reversible process, namely of how the system may be restored to the initial state reversibly.

Since the notion of reversibility is global in time, one expects that the restriction of a reversible process P of duration d to an interval $[0, \bar{d})$, $\bar{d} < d$ to be reversible, too. Such is really the case.

PROPOSITION 2.4.1. *Any restriction of a reversible process P is a reversible process.*

The property (b) is obviously true for the restriction. As for (a), it suffices to observe that since $\tilde{P}$ is reversible on $[0, d)$ then

$$\varrho(\sigma, P_{\bar{d}}) = \varrho(\varrho(\sigma, P), \tilde{P}_{d-\bar{d}})$$

and

$$\varrho(\sigma, P_t) = \varrho(\varrho(\varrho(\sigma, P), \tilde{P}_{d-\bar{d}}), \tilde{P}_{\bar{d}-t}), \qquad t \in [0, \bar{d}).$$

Let

$$\Gamma(\sigma, P) := \int_0^d \hat{\gamma}(\sigma(t), P(t))dt$$

be the *global dissipation* along the process P. The next theorem provides a fundamental property of reversible processes.

THEOREM 2.4.1. *If the global dissipation $\Gamma(\sigma, P)$ has values in $\mathbb{R}^-$ for any process P, then it vanishes at any reversible process.*

Proof. Let P be any reversible process. By (a) we have

$$\int_0^d \hat{\gamma}(\tilde{\sigma}(t), \tilde{P}(t))dt = \int_0^d \hat{\gamma}(\sigma(d-t), \tilde{P}(t))dt$$

and by (b)

$$\int_0^d \hat{\gamma}(\sigma(d-t), \tilde{P}(t))dt = -\int_0^d \hat{\gamma}(\sigma(d-t), P(d-t))dt.$$

Then a change of variable yields

$$\Gamma(\varrho(\sigma, P), \tilde{P}) = -\Gamma(\sigma, P).$$

The hypothesis that Γ is negative-valued for any process gives the desired conclusion. $\square$

In the next section we reinvestigate the nature of the second law in terms of reversible and irreversible processes. Before that, it seems worth examining whether the reversibility property holds for processes in some customary models of continua.

2.4.1. Thermoelastic solid. With reference to (2.1.5), thermal and mechanical effects are considered by letting σ be the pair

$$\sigma = (\mathbf{F}, \theta),$$

namely the configuration C. Then $\Sigma = \mathrm{Lin}^+ \times \mathbb{R}^{++}$. Letting $P = (\mathbf{L}, \dot{\theta}, \mathbf{g})$ be a given process on $[0, d)$, we have the current state $\sigma(t)$ as discussed in §2.2. To examine the property (a) let $\sigma_0 = (\mathbf{F}_0, \theta_0)$ and denote by $\sigma_d = (\mathbf{F}_d, \theta_d)$ the value of σ at $t = d$. In $[0, d)$ we have

$$\varrho(\sigma_0, P_t) = \left(\mathbf{F}_0 + \int_0^t \mathbf{L}(\xi)\mathbf{F}(\xi)\, d\xi, \quad \theta_0 + \int_0^t \dot{\theta}(\xi)\, d\xi\right).$$

Because $\tilde{P}(\xi) = (-\dot{\theta}(d-\xi), -\dot{\mathbf{L}}(d-\xi), -\mathbf{g}(d-\xi))$ we have

$$\varrho(\sigma_d, \tilde{P}_{d-t}) = \left(\mathbf{F}_d - \int_d^{d-t} \mathbf{L}(d-\xi)\, \mathbf{F}(d-\xi)\, d\xi, \quad \theta_d + \int_d^{d-t} \frac{d\theta(d-\xi)}{d\xi} d\xi\right).$$

Then

$$\varrho(\sigma_d, \tilde{P}_{d-t}) = \left(\mathbf{F}_d + \int_0^t \mathbf{L}(\zeta)\, \mathbf{F}(\zeta)\, d\zeta, \quad \theta_d + \theta(t) - \theta_0\right).$$

Accordingly, the property (a) requires that

$$\mathbf{F}_d = \mathbf{F}_0, \qquad \theta_d = \theta_0,$$

namely the pair (σ, P) is required to be cyclic.

In reference to (b), observe that ϵ and $\mathbf{T}$ are functions of σ and then

$$\gamma_1 := \frac{1}{\theta}\dot{\epsilon} - \frac{1}{\rho\theta}\mathbf{T} \cdot \mathbf{L}$$

is a function of $\sigma, \mathbf{L}, \dot{\theta}$ which depends linearly on $\mathbf{L}$ and $\dot{\theta}$. Hence it follows at once that γ_1 meets (b). Since

$$\gamma = \gamma_1 + \frac{1}{\rho\theta}\mathbf{q} \cdot \mathbf{g},$$

(b) requires that

$$\hat{\mathbf{q}}(\sigma(t), \mathbf{g}(t)) \cdot \mathbf{g}(t) = -\hat{\mathbf{q}}(\tilde{\sigma}(d-t), \tilde{\mathbf{g}}(d-t)) \cdot \tilde{\mathbf{g}}(d-t)$$

or, because σ is even and $\mathbf{g}$ is odd,

$$\hat{\mathbf{q}}(\sigma(t), \mathbf{g}(t)) \cdot \mathbf{g}(t) = \hat{\mathbf{q}}(\sigma(t), -\mathbf{g}(t)) \cdot \mathbf{g}(t),$$

which means that $\mathbf{q} \cdot \mathbf{g} = 0$ in $[0, d)$, namely

$$\mathbf{g}(t) = 0, \qquad t \in [0, d).$$

Then the set of reversible processes consists of the processes $P = (\mathbf{L}, \dot{\theta}, 0)$.

2.4.2. Newtonian viscous fluid. As with the thermoelastic solid, according to (2.1.7) the state of the viscous fluid is the pair

$$\sigma = (\mathbf{F}, \theta).$$

The stress tensor $\mathbf{T}$ and the heat flux $\mathbf{q}$ depend on the state σ and the present value of the process P through $\mathbf{L}$ and $\mathbf{g}$. Again the property (a) is obviously satisfied. For the property (b) observe that

$$\gamma = \gamma_1 + \gamma_2$$

where

$$\gamma_1 = \frac{1}{\theta}\dot{\epsilon} + \frac{p}{\rho\theta}\mathrm{tr}\mathbf{D}, \qquad \gamma_2 = -\frac{1}{\rho\theta}[2\mu\mathbf{D} \cdot \mathbf{D} + \lambda(\mathrm{tr}\mathbf{D})^2] + \frac{1}{\rho\theta^2}\mathbf{g} \cdot \mathbf{K}\mathbf{g};$$

the quantities μ, λ, and $\mathbf{K}$ are supposed to be constant and to be such that $\mu > 0$, $3\lambda + 2\mu > 0$ and $\mathbf{K}$ is positive definite.

The dissipation γ_1 is linear in $\mathbf{L}, \dot{\theta}$ and then complies with (b). Being quadratic in $\mathbf{L}$, γ_2 is even under time reversal; compatibility with (b) requires that $\gamma_2 = 0$, whence

$$\mathbf{D} = 0, \qquad \mathbf{g} = 0 \quad \text{in } [0, d).$$

So, apart from the trivial process, in (Newtonian) viscous fluids, processes are irreversible.

2.4.3. Material with memory. Consider the material with fading memory described by (2.1.8). The state

$$\sigma = (\mathbf{F}^t, \theta^t)$$

is the pair of the history of the deformation gradient and the history of the temperature. The process

$$P = (\mathbf{L}, \dot{\theta}, \mathbf{g})$$

is the usual triplet. In the previous models property (a) is trivially true. Here, because of the memory, (a) is not satisfied since (a) demands, in particular, that

$$\theta^t = \theta^d - \int_d^{d-t} \dot{\theta}(\xi) d\xi,$$

namely

$$\theta(t-s) = \theta(d-s) - \theta(d-t) + \theta(d), \qquad s \in \mathbb{R}^+, \qquad t \in [0, d).$$

The arbitrariness of s and t implies that (a) holds only if θ is constant in time. By the same token we conclude that (a) holds only if $\mathbf{F}$ is constant in time. As before, we see also that (b) holds only if $\mathbf{q} \cdot \mathbf{g} = 0$. So, in general, for time-dependent processes memory results in irreversibility.

Remark 2.4.1. Equation (2.4.1) might be replaced by

$$(2.4.1') \qquad \tilde{P}(t) = (\tilde{\mathbf{L}}(t), \dot{\tilde{\theta}}(t), \tilde{\mathbf{g}}(t)) = (-\mathbf{L}(d-t), -\dot{\theta}(d-t), \mathbf{g}(d-t)),$$

which is favoured by the second author. He has in mind the picture that the position vector and the temperature are invariant in passing from a process to its reverse; time derivatives change sign whereas space derivatives do not. For the customary models of continua examined here, equations (2.4.1) and (2.4.1′) lead to almost the same conclusions and that is why the definition of $\tilde{\mathbf{g}}$ may appear questionable.

2.5. A more restrictive statement of the second law. As to the second law, the detailed analysis about reversibility and irreversibility performed in standard textbooks leads ultimately to a formulation of the second law through the principle of the increase of entropy. The key property within this framework is usually stated as follows.

With reference to integrals along cycles and letting dQ be the "infinitesimal amount of heat" received by the body from a source at the temperature θ, it is assumed that

$$\int \frac{dQ}{\theta} \leq 0$$

is valid for all cycles, the equality sign referring to reversible processes and the inequality to irreversible processes.

Now this view is extended to continuum thermodynamics in the sense that the equality sign in the statement of the second law (or entropy inequality) is taken to hold for reversible processes only. According to Theorem 2.4.1, the dissipation $\Gamma(\sigma, P)$ vanishes at the reversible process. Then the statement of the second law, provided in §2.2, becomes better in the following way.

Second law. *For every cycle* $(\sigma, P) \in \Sigma \Diamond \Pi$ *the inequality*

$$\int_0^{d_P} \left[\frac{\hat{h}(\sigma(t), P(t))}{\theta(t)} + \frac{1}{\rho(t)\theta^2(t)} \hat{\mathbf{q}}(\sigma(t), P(t)) \cdot \mathbf{g}(t) \right] dt \leq 0$$

holds with $\sigma(t) = \varrho(\sigma, P_t)$*; the equality sign holds if and only if the process is reversible.*

2.6. Work and dissipativity. In the fifties and sixties, especially before the development of rational thermodynamics, remarkable attention was devoted to a property which, in a sense, was regarded as a condition of thermodynamic character. Such a property is named *dissipativity* and can be traced back to Drucker and König and Meixner [88]. Both for completeness about thermodynamic aspects and for future reference, it is worth recalling the statement.

As a preliminary remark, observe that, by the energy balance in the form (2.2.6), $\mathbf{T} \cdot \mathbf{L}$ is the power, per unit volume, acting *on* the body through the stress forces. The corresponding work in a time interval is just the integral in time of $\mathbf{T} \cdot \mathbf{L}$. Let $\mathbf{S}$ be the first Piola-Kirchhoff stress tensor (cf. [121], §43A), namely

$$\mathbf{S} = J\,\mathbf{T}(\mathbf{F}^T)^{-1}.$$

Since $\mathbf{L} = \dot{\mathbf{F}}\,\mathbf{F}^{-1}$ then

$$\mathbf{S} \cdot \dot{\mathbf{F}} = J\,\mathbf{T} \cdot \mathbf{L}$$

is the power of the stress, per unit volume, in the reference placement.

Disregarding thermal effects, we assume that $\mathbf{S}$ is given by a suitable functional of the history of $\mathbf{F}$.

For any material point $\mathbf{X}$ in the body, consider (deformation gradient) functions $\mathbf{F} : \mathbb{R} \to \mathrm{Lin}^+$ whose derivative has compact support. These functions are sometimes called *paths.* For formal simplicity we let $\dot{\mathbf{F}} = 0$ up to $t = 0$. By following the nomenclature on the subject we call *virgin state* the zero deformation gradient.

Dissipativity means that *the work*

$$w(\mathbf{F}^t) = \int_0^t \mathbf{S}(\mathbf{F}^\tau) \cdot \dot{\mathbf{F}}(\tau)\, d\tau$$

is positive for any path starting from the virgin state and any time $t \in \mathbb{R}^+$. In other words, the assumption is that work must be done to deform a solid from its virgin state.

Consequences of dissipativity were derived by Gurtin and Herrera [77] in the case of linear viscoelasticity. Their results have been reexamined by Day [37], [38] and contrasted with the consequences of the thermodynamic assumption that the work is positive for every closed path.[7] In the next chapter we

[7] Closed means that $\mathbf{F}(-\infty) = \mathbf{F}(\infty)$.

develop a thermodynamic analysis of (linear) viscoelasticity; in that context we review the results obtained by Gurtin and Herrera and Day.

2.7. Exploitation of the Clausius-Duhem inequality. In order to outline a standard procedure in the thermodynamics of materials with memory and to establish a precise relationship among the Clausius inequality and the Clausius-Duhem inequality, now we examine the restrictions placed on a material with memory by the second law as expressed by the Clausius-Duhem inequality.

For the sake of generality consider a material whose state σ is the triplet $(\mathbf{F}^t, \theta^t, \mathbf{g}^t)$ of the histories of $\mathbf{F}, \theta$, and $\mathbf{g}$. The response of the material is characterized by the functionals $\bar{\psi}, \bar{\eta}, \bar{\mathbf{S}}, \bar{\mathbf{q}}$ for the free energy, the entropy, the second Piola-Kirchhoff stress tensor, and the heat flux. Let $\mathbf{F}, \theta, \mathbf{g}$ on $\mathbb{R}$ be mild functions and let $\bar{\eta}, \bar{\mathbf{S}}, \bar{\mathbf{q}}$ be continuous functionals while $\bar{\psi}$ is continuously differentiable. Then the Clausius-Duhem inequality in the form (1.4.9), or rather

$$-(\dot{\psi} + \eta\dot{\theta}) + \frac{1}{\rho_0}\mathbf{S}\cdot\dot{\mathbf{F}} - \frac{1}{\rho\theta}\mathbf{q}\cdot\mathbf{g} \geq 0,$$

becomes

(2.7.1)

$$\left[\frac{\partial\bar{\psi}}{\partial\mathbf{F}}(\sigma(t)) - \frac{1}{\rho_0}\bar{\mathbf{S}}(\sigma(t))\right]\cdot\dot{\mathbf{F}}(t) + \left[\frac{\partial\bar{\psi}}{\partial\theta}(\sigma(t)) + \bar{\eta}(\sigma(t))\right]\dot{\theta}(t)$$
$$+\frac{\partial\bar{\psi}}{\partial\mathbf{g}}(\sigma(t))\cdot\dot{\mathbf{g}}(t) + d\bar{\psi}(\sigma(t)|\dot{\Lambda}^t) + \frac{1}{\rho(t)\,\theta(t)}\mathbf{g}(t)\cdot\bar{\mathbf{q}}(\sigma(t)) \leq 0,$$

where $\Lambda^t = ({}_r\mathbf{F}^t,\, {}_r\theta^t,\, {}_r\mathbf{g}^t)$. The second law of thermodynamics requires that (2.7.1) be true for every set of (admissible) histories $\sigma(t)$, which places restrictions on the constitutive functionals $\bar{\psi}, \bar{\eta}, \bar{\mathbf{S}}, \bar{\mathbf{q}}$. To derive such restrictions we need the following property.

LEMMA 2.7.1. *Let $\phi(\cdot) : \mathbb{R} \to W$ be a mild function, t any time, and ω any element of W. Then there are functions $\phi_\alpha(\cdot) : \mathbb{R} \to W, \alpha \in \mathbb{R}^{++}$, likewise meeting the mildness property, such that*

$$\phi_\alpha(t) = \phi(t), \qquad \dot{\phi}_\alpha(t) = \omega, \qquad \|\phi_\alpha^t - \phi^t\| = O(\alpha), \qquad \|\dot{\phi}_\alpha^t - \dot{\phi}^t\| = O(\alpha).$$

Proof (Day [38]). Let $p \in C^2(\mathbb{R}, \mathbb{R})$ such that

$$p(0) = 0, \qquad \dot{p}(0) = 0, \qquad p(\xi) = 0 \quad \text{as} \quad |\xi| \geq 1;$$

for example, $p(\xi) = 0$ as $|\xi| \geq 1$ and $p(\xi) = \xi(1-\xi^2)^3$ as $|\xi| \leq 1$ works. Choose

$$\phi_\alpha(\xi) = \phi(\xi) + \alpha\, p\Big(\frac{\xi - t}{\alpha}\Big)\,(\omega - \dot{\phi}(t)).$$

Obviously, $\phi_\alpha(t) = \phi(t)$, $\dot{\phi}_\alpha(t) = \omega$. Since the support of $p, \dot{p}, \ddot{p}$ is the interval $(t-\alpha, t+\alpha)$, it follows that

$$\|\phi_\alpha^t - \phi^t\| = O(\alpha), \quad \|\dot{\phi}_\alpha^t - \dot{\phi}^t\| = O(\alpha), \quad \|\ddot{\phi}_\alpha^t - \ddot{\phi}^t\| = O(\alpha). \qquad \square$$

Incidentally, admissible histories ϕ^t might be subject to constraints, usually in the form of strict inequalities.[8] The choice of a sufficiently small α allows ϕ_α^t to satisfy the constraints whenever ϕ^t does.

Look at histories $\mathbf{F}_\alpha^t, \theta_\alpha^t, \mathbf{g}_\alpha^t$, in the sense of Lemma 2.7.1, such that

$$\dot{\mathbf{F}}_\alpha(t) = \mathbf{A}, \qquad \dot{\theta}_\alpha(t) = a, \qquad \dot{\mathbf{g}}_\alpha(t) = \mathbf{a},$$

where $\mathbf{A} \in \mathrm{Lin}$, $a \in \mathbb{R}$, $\mathbf{a} \in V$. The inequality (2.7.1) must hold in connection with $\mathbf{F}_\alpha^t, \theta_\alpha^t, \mathbf{g}_\alpha^t$ as well; replacement of $\mathbf{F}^t, \theta^t, \mathbf{g}^t$ with $\mathbf{F}_\alpha^t, \theta_\alpha^t, \mathbf{g}_\alpha^t$ in (2.7.1) and the limit $\alpha \to 0$ give

$$\left[\frac{\partial\bar{\psi}}{\partial\mathbf{F}}(\sigma(t)) - \frac{1}{\rho_0}\bar{\mathbf{S}}(\sigma(t))\right] \cdot \mathbf{A} + \left[\frac{\partial\bar{\psi}}{\partial\theta}(\sigma(t)) + \bar{\eta}(\sigma(t))\right] a + \frac{\partial\bar{\psi}}{\partial\mathbf{g}}(\sigma(t)) \cdot \mathbf{a}$$

$$+ d\bar{\psi}(\sigma(t)|\dot{\Lambda}^t) + \frac{1}{\rho(t)\,\theta(t)}\mathbf{g}(t) \cdot \bar{\mathbf{q}}(\sigma(t)) \le 0.$$

The arbitrariness of $\mathbf{A}, a$, and $\mathbf{a}$ implies that

$$\frac{\partial\bar{\psi}}{\partial\mathbf{g}} = 0; \tag{2.7.2}$$

the functionals $\bar{\eta}, \bar{\mathbf{S}}, \bar{\psi}$ are related by

$$\bar{\eta} = -\frac{\partial\bar{\psi}}{\partial\theta}, \tag{2.7.3}$$

$$\bar{\mathbf{S}} = \rho_0 \frac{\partial\bar{\psi}}{\partial\mathbf{F}}, \tag{2.7.4}$$

and the *generalised dissipation inequality*

$$d\bar{\psi} + \frac{1}{\rho\theta}\bar{\mathbf{q}} \cdot \mathbf{g} \le 0 \tag{2.7.5}$$

holds. We see at once that (2.7.2)–(2.7.5) are also sufficient for the Clausius-Duhem inequality (2.7.1) to hold identically. This constitutes the proof of the following theorem.

---

[8]Typically, $\phi > 0$ for the absolute temperature and $\det\phi > 0$ for the deformation gradient.

THEOREM 2.7.1. *The functionals $\bar{\psi}, \bar{\eta}, \bar{\mathbf{S}}, \bar{\mathbf{q}}$ are compatible with the second law of thermodynamics, in the form of the Clausius–Duhem inequality, if and only if* (2.7.2)–(2.7.5) *hold.*[9]

Incidentally, for future reference we observe that by (2.7.4) the Cauchy stress tensor is given by

$$\bar{\mathbf{T}} = \frac{\rho_0}{J} \frac{\partial \bar{\psi}}{\partial \mathbf{F}} \mathbf{F}^T. \tag{2.7.6}$$

As a comment to Theorem 2.7.1 we remark that the stress tensor, both $\bar{\mathbf{S}}$ and $\bar{\mathbf{T}}$, and the entropy $\bar{\eta}$, along with the free energy $\bar{\psi}$ are independent of the present value of the temperature gradient. Thermodynamically, however, we cannot prevent the response functionals from depending on the past history ${}_r\mathbf{g}^t$. The following property extends to functionals which depend on ${}_r\mathbf{g}^t$, a statement which is usually referred to functionals independent of ${}_r\mathbf{g}^t$.

PROPOSITION 2.7.1 (cf. [100], §3.3). *Among all histories $\mathbf{F}^t, \theta^t, \mathbf{g}^t$, with given present values $\mathbf{F}_0, \theta_0, \mathbf{0}$, none yields a smaller value of the free energy than that corresponding to the constant histories $\mathbf{F}_0^\dagger, \theta_0^\dagger, \mathbf{0}^\dagger$.*

Proof. By (2.7.5) it follows that $d\bar{\psi}$ is negative for all (admissible) histories of $\mathbf{g}$ with a vanishing present value. Given arbitrary, but fixed, histories $\mathbf{F}^{t_0}, \theta^{t_0}, \mathbf{g}^{t_0}$, with $\mathbf{g}(t_0) = 0$, consider their static continuation up to time $t > t_0$. For any value of $t > t_0$ we have

$$\dot{\psi}(t) = d\bar{\psi} \leq 0.$$

Let $\Lambda_0 = (\mathbf{F}(t_0), \theta(t_0), \mathbf{0})$. Accordingly

$$\bar{\psi}(\Lambda_0, \Lambda^t) \leq \bar{\psi}(\Lambda_0, \Lambda^{t_0}).$$

Because of the property (1.2.2) of static continuations and the continuity of the functional $\bar{\psi}$, we have

$$\lim_{t \to \infty} \bar{\psi}(\Lambda_0, \Lambda^t) = \bar{\psi}(\Lambda_0, \Lambda_0^\dagger).$$

Hence

$$\bar{\psi}(\Lambda_0, \Lambda_0^\dagger) \leq \bar{\psi}(\Lambda_0, \Lambda^{t_0}). \tag{2.7.7}$$ □

2.8. Clausius inequality. There are some objections against using the Clausius-Duhem inequality as an expression of the second law. First, it is assumed at the outset that an entropy functional, as well as a free energy

[9]The results contained in this theorem were derived by Coleman [19] in the instance when the functionals $\bar{\psi}, \bar{\eta}, \bar{\mathbf{S}}, \bar{\mathbf{q}}$ are taken at the outset as dependent on the temperature gradient through the present value only.

functional, exists. However, as shown by Day [40] in the case of rigid heat conductors, really there exists a family of entropy functionals, all compatible with the Clausius-Duhem inequality. Thus it seems more appropriate to consider statements of the second law where entropy and free energy do not enter the theory at the very beginning but, hopefully, are derived as a consequence of the thermodynamic theory. Such is just the case of the other statements of the second law examined before. It goes without saying that, for materials with memory, the evaluation of the free energy and entropy functionals is certainly a nontrivial task and may be plagued with nonuniqueness. These aspects are investigated in detail in conjunction with linear viscoelasticity.

Second, the Clausius-Duhem inequality appears to be overly restrictive in that it is a local-in-time condition rather than a condition for a cycle (or an approximate cycle) as a whole. That is why later developments are based on global statements for the second law. Yet there are cases where the Clausius-Duhem inequality is equivalent to global statements like, e.g., the Clausius inequality. Whether the Clausius-Duhem inequality is more restrictive or not depends on the form of constitutive functionals, as will be shown.

We assume that the state of the material σ is the triplet $(\mathbf{F}^t, \epsilon^t, \mathbf{g}^t)$ and $\eta, \theta, \mathbf{S}, \mathbf{q}$ are determined by σ through the functionals $\breve{\eta}, \breve{\theta}, \breve{\mathbf{S}}, \breve{\mathbf{q}}$. Then the Clausius-Duhem inequality, which can be given the form

$$\dot{\eta} \geq \frac{1}{\theta}\dot{\epsilon} - \frac{1}{\rho_0 \theta}\mathbf{S}\cdot\dot{\mathbf{F}} + \frac{1}{\rho\theta^2}\mathbf{q}\cdot\mathbf{g},$$

leads to

$$\frac{\partial \breve{\eta}}{\partial \mathbf{g}} = 0, \qquad \breve{\theta} = 1 \Big/ \frac{\partial \breve{\eta}}{\partial \epsilon}, \qquad \breve{\mathbf{S}} = -\rho_0 \frac{\partial \breve{\eta}}{\partial \mathbf{F}} \Big/ \frac{\partial \breve{\eta}}{\partial \epsilon}, \qquad d\breve{\eta} - \frac{1}{\rho\theta^2}\breve{\mathbf{q}}\cdot\mathbf{g} \geq 0.$$

Then, by use of (1.2.2) and paralleling the proof of Proposition 2.7.1, we can show that, for any set of histories $\Gamma^t = (\mathbf{F}^t, \epsilon^t, \mathbf{g}^t)$ such that $\Gamma(t_0) = \Gamma_0 := (\mathbf{F}_0, \epsilon_0, \mathbf{0})$, $\mathbf{F}_0 \in \mathrm{Lin}^+, \epsilon_0 \in \mathbb{R}$, we have

$$\breve{\eta}(\Gamma_0, \Gamma_0^\dagger) \geq \breve{\eta}(\Gamma_0, \Gamma^{t_0}). \tag{2.8.1}$$

These results allow us to prove that the validity of the Clausius-Duhem inequality implies that of the Clausius inequality as follows.

THEOREM 2.8.1. *If the material is specified by the functionals* $\breve{\eta}, \breve{\theta}, \breve{\mathbf{S}}, \breve{\mathbf{q}}$ *then the validity of the Clausius-Duhem inequality implies that of the Clausius inequality for every process, in* $[t_0, t_1)$, *such that* $\Gamma(t_1) = \Gamma(t_0) = (\mathbf{F}_0, \epsilon_0, \mathbf{0})$, $\mathbf{F}_0 \in \mathrm{Lin}^+, \epsilon_0 \in \mathbb{R}$.

Proof. Consider two times t_0, t_1 with $t_1 > t_0$. Integration of (1.4.8) yields

$$\int_{t_0}^{t_1} \left[-\frac{1}{\rho}\nabla\cdot\left(\frac{1}{\theta}\mathbf{q}\right) + \frac{1}{\theta}r \right] dt \;\leq\; \eta(t_1) - \eta(t_0).$$

Let $\mathbf{F}(\cdot)$, $\epsilon(\cdot)$ be constant and $\mathbf{g}(\cdot)$ vanish at all times prior to time t_0. Then for any set of histories Γ^t with $\mathbf{F}(t_1) = \mathbf{F}(t_0)$, $\epsilon(t_1) = \epsilon(t_0)$, $\mathbf{g}(t_1) = \mathbf{g}(t_0) = 0$, (2.8.1) yields

$$\eta(t_1) \leq \eta(t_0).$$

Hence we recover the Clausius inequality

$$\int_{t_0}^{t_1} \left[-\frac{1}{\rho} \nabla \cdot \left(\frac{1}{\theta} \mathbf{q} \right) + \frac{1}{\theta} r \right] dt \leq 0$$

for any set of histories Γ^t with $\Gamma(t_1) = \Gamma(t_0) = (\mathbf{F}_0, \epsilon_0, \mathbf{0})$, $\mathbf{F}_0 \in \mathrm{Lin}^+, \epsilon_0 \in \mathbb{R}$. □

The converse does not appear to be always true also because in a theory based on the Clausius inequality the functionals for η and ψ must be derived and are not supposed to be given at the outset. For some materials, though, we can prove that the Clausius-Duhem inequality holds as a consequence of the Clausius inequality. This is the case in the following constitutive scheme. The heat flux is given by the relation

$$\mathbf{q}(t) = \mathbf{q}(\mathbf{F}(t), \epsilon(t), \mathbf{g}(t))$$

whereby only the present value of $\mathbf{F}, \epsilon, \mathbf{g}$ occurs. Memory effects are allowed as to $\mathbf{S}$ and θ in the form

$$\mathbf{S}(t) = \check{\mathbf{S}}(\mathbf{F}^t, \epsilon^t), \qquad \theta(t) = \check{\theta}(\mathbf{F}^t, \epsilon^t).$$

Then, as shown by Day [38] and Coleman and Owen [29], there exists the entropy functional $\check{\eta}(\mathbf{F}^t, \theta^t)$ which satisfies the Clausius-Planck inequality

$$\dot{\eta} \geq \frac{1}{\theta} \left(\dot{\epsilon} - \frac{1}{\rho_0} \mathbf{S} \cdot \dot{\mathbf{F}} \right)$$

while the heat flux satisfies the heat conduction inequality

$$0 \geq \frac{1}{\rho \theta^2} \mathbf{q} \cdot \mathbf{g}.$$

The addition of these two inequalities yields

$$\dot{\eta} \geq \frac{1}{\theta} \left(\dot{\epsilon} - \frac{1}{\rho_0} \mathbf{S} \cdot \dot{\mathbf{F}} \right) + \frac{1}{\rho \theta^2} \mathbf{q} \cdot \mathbf{g},$$

namely the Clausius-Duhem inequality.

So, in general, a thermodynamic theory based on the Clausius inequality, or other global statements, is less restrictive than the theory based on the Clausius-Duhem inequality.

Chapter 3

Linear Viscoelasticity

3.1. The linear viscoelastic solid. Linear (infinitesimal) viscoelasticity and especially the model of linear viscoelastic solid can be traced back to Boltzmann [7] who, roughly speaking, considered an elastic material with memory. He elaborated the model of (linear) viscoelastic solid on the basis of the following assumptions.

At any (fixed) point $\mathbf{X}$ of the body, the stress at any time t depends upon the strain at all preceding times. If the strain at all preceding times is in the same direction, then the effect is to reduce the corresponding stress. The influence of a previous strain on the stress depends on the time elapsed since that strain occurred and is weaker for those strains that occurred long ago. Such properties make the model of solid elaborated by Boltzmann a material with (fading) memory. In addition, Boltzmann made the assumption that a superposition of the influence of previous strains holds, which means that the stress-strain relation is linear.

This is the essence of the linear viscoelastic solid. Quite often linear viscoelasticity is motivated through elementary models of the solid element in terms of series of springs and dash-pots (cf. [5]). Rather, we prefer to begin the analysis by framing the model of linear viscoelasticity within the scheme of materials with fading memory.

To be specific, we consider viscoelasticity in the isothermal approximation, which means that the temperature does not enter the model (state and constitutive relation). So the state involves the deformation gradient only while the constitutive equation is in fact a stress-strain relation. Consistent with the superposition principle and the underlying linearity assumption, we consider a displacement field $\mathbf{u}(\mathbf{X},t)$ subject to

$$\sup_{\mathbf{X},t}\left|\frac{\partial \mathbf{u}}{\partial \mathbf{X}}(\mathbf{X},t)\right| \ll 1.$$

Then we approximate the strain tensor $\mathbf{E}=\frac{1}{2}(\mathbf{C}-\mathbf{1})$ with the linearized strain tensor sym $\partial \mathbf{u}/\partial \mathbf{X}$. The state of the body, σ, is then taken as the history of $\mathbf{E}$, viz.

$$\sigma(t) = (\mathbf{E}(t),\, {}_r\mathbf{E}^t).$$

Owing to the linear approximation, we identify the dependence on the reference position $\mathbf{X}$ with the dependence on the present position $\mathbf{x}$. Hence we let $\mathbf{x} \in \mathcal{R}$ and identify $\mathbf{E}$ with the *infinitesimal strain tensor,* namely

$$\mathbf{E} = \operatorname{sym} \frac{\partial \mathbf{u}}{\partial \mathbf{x}}.$$

Accordingly we let $J = 1$ and then ρ be constant in time; ρ is a given function of the position $\mathbf{x} \in \mathcal{R}$. Moreover we identify the (first) Piola-Kirchhoff stress tensor $\mathbf{S}$ with the Cauchy stress tensor $\mathbf{T}$ and the power of the stress $\mathbf{T} \cdot \mathbf{D}$ (or $\mathbf{S} \cdot \dot{\mathbf{F}}$) can then be identified with $\mathbf{T} \cdot \dot{\mathbf{E}}$.

With this in mind, letting $k \in L^1(\mathbb{R}^+)$ we consider the fading memory space

$$\mathcal{H} = \left\{ \mathbf{E}^t : \mathbb{R}^+ \to \mathrm{Sym}, \quad \int_0^\infty |{}_r\mathbf{E}^t(s)|^2 k(s) ds < \infty \right\}$$

with the norm (1.3.1) and the associated inner product (cf. §1.3). Once we make the assumption that the constitutive equation for the stress $\mathbf{T}$ is given by a bounded, linear functional on $\mathcal{H}$, then the Riesz theorem gives the representation (1.3.4), namely

$$\boldsymbol{\mathcal{T}}(\mathbf{E}^t) = \boldsymbol{\mathcal{G}}(0)\, \mathbf{E}(t) + \int_0^\infty \boldsymbol{\mathcal{G}}(s)\, \mathbf{E}(t-s)\, ds,$$

where now $\boldsymbol{\mathcal{G}}$ has values in the space of fourth-order tensors Lin(Sym), for any $s \in \mathbb{R}^+$. Of course, $\boldsymbol{\mathcal{G}}(0)$ takes finite values and $\boldsymbol{\mathcal{G}}$ has a finite L^2 norm (cf. (1.3.4) and (1.3.5)) with weight k^{-1}. This relation makes it apparent how the stress $\mathbf{T}(t)$ depends on both the present value $\mathbf{E}(t)$ and the (past) history ${}_r\mathbf{E}^t$. The function $\boldsymbol{\mathcal{G}}(s)$ is sometimes called *Boltzmann function* and is the kernel of the linear functional $\boldsymbol{\mathcal{T}}$ for $\mathbf{T}$. To adhere to the standard notation let $\mathbf{G} : \mathbb{R}^+ \to \mathrm{Lin(Sym)}$ be such that

$$\mathbf{G}'(s) = \boldsymbol{\mathcal{G}}(s), \qquad \mathbf{G}(0) = \boldsymbol{\mathcal{G}}(0),$$

the prime denoting the derivative with respect to s. The solution

$$\mathbf{G}(s) = \mathbf{G}(0) + \int_0^s \boldsymbol{\mathcal{G}}(\xi)\, d\xi$$

is called *relaxation function.* Of course, $\mathbf{G}$ is allowed to depend also on the position $\mathbf{x}$ but, as usual, such dependence will be understood and not written.

Because $\mathcal{G}k^{-1} \in \mathcal{H}$ we have

$$|\mathbf{G}(0)| < \infty, \qquad \int_0^\infty |\mathbf{G}'(s)|^2 k^{-1}(s)\, ds < \infty,$$

which is usually assumed in linear viscoelasticity. Now observe that since $|\mathbf{G}'|^2 k^{-1}$, $k \in L^1(\mathbb{R}^+)$, then by the Cauchy-Schwarz inequality we have

$$\left(\int_0^\infty |\mathbf{G}'(s)|\, ds\right)^2 \leq \int_0^\infty |\mathbf{G}'(s)|^2\, k^{-1}(s)\, ds \quad \int_0^\infty k(s)\, ds$$

and then $\mathbf{G}' \in L^1(\mathbb{R}^+)$. This in turn makes it evident that $\mathcal{T}$ is a bounded, linear functional on $\mathcal{H}$.

There are cases where this scheme is much too restrictive; $\mathbf{G}'(0)$ may be unbounded (and $\mathbf{G}'$ nonintegrable on $\mathbb{R}^+$) and $\mathbf{G}(0)$ may be unbounded too. Such cases are confined to Chapter 7.

The fourth-order tensor $\mathbf{G}_0 := \mathbf{G}(0)$ is called an *instantaneous elastic modulus* and governs the response to instantaneous changes in strain. The limit of $\mathbf{G}(s)$, as $s \to \infty$, is supposed to exist and

$$\mathbf{G}_\infty = \lim_{s\to\infty} \mathbf{G}(s)$$

is called an *equilibrium elastic modulus* in that the constancy of the history $\mathbf{E}^t$, $\mathbf{E}^t(s) = \mathbf{E}, s \in \mathbb{R}^+$, gives

$$\mathbf{T}(t) = \mathbf{G}_\infty\, \mathbf{E}.$$

In a solid, a nonzero, constant strain is supposed to induce a nonzero stress such that $\mathbf{T}\cdot\mathbf{E}$ is strictly positive. This is expressed by letting[10]

$$\mathbf{G}_\infty > 0. \tag{3.1.1}$$

For later convenience we write the constitutive equation of linear viscoelasticity as

$$\mathbf{T}(t) = \mathbf{G}_0\, \mathbf{E}(t) + \int_0^\infty \mathbf{G}'(s)\mathbf{E}(t-s)\, ds. \tag{3.1.2}$$

If the strain function has compact support, whence $\mathbf{E}(-\infty) = 0$, then an obvious integration by parts yields

$$\mathbf{T}(t) = \int_0^\infty \mathbf{G}(s)\, \dot{\mathbf{E}}(t-s)\, ds, \tag{3.1.3}$$

[10]For any tensor $\mathbf{A}$ the writing $\mathbf{A} > 0$ means that $\mathbf{A}$ is positive definite. The symmetry of $\mathbf{G}_\infty$ is in fact a thermodynamic restriction.

a superposed dot denoting time differentiation. Incidentally, (3.1.3) motivates why $\mathbf{G}$, rather than $\mathbf{G}'$, is called a relaxation function.

Remark 3.1.1. Although outstanding improvements in the theory of linear viscoelasticity have been obtained in the last 30 years, the early contributions are due to such scientists as Maxwell, Kelvin, and Voigt. Their names are associated with two elementary models of dissipative solids in terms of a spring and a dash-pot in series or in parallel (Maxwell's and Kelvin–Voigt's). Boltzmann [7] in 1874 was the first to develop a three-dimensional theory of isotropic viscoelasticity. Later Volterra [125]–[128] obtained comparable results for anisotropic solids.

Remark 3.1.2. About the nomenclature, Maxwell [96] was the first to use the term "historical" to describe the feature that the stress at a given time depends not only on the strain at that time but also on the history of the strain up to that time. In Volterra's nomenclature [128], [129], such behaviours are regarded as "hereditary phenomena" and "heredity" is the response of the material due to the (past) history. So, with reference to (3.1.2), the integral gives the heredity of the strain.

It may be convenient (cf. §6.3) to model the viscoelastic solid as a material with internal variables [22], [100]. In such a case, letting the body be isotropic we write the analogue of (3.1.2) as

$$\mathbf{T}(t) = \beta_0(\mathrm{tr}\,\mathbf{E})\mathbf{1} + 2\mu_0 \overset{\circ}{\mathbf{E}} + \boldsymbol{\xi}$$

where the internal variable $\boldsymbol{\xi} \in \mathrm{Sym}$ is given by the growth equation

$$\dot{\boldsymbol{\xi}} = -\alpha\boldsymbol{\xi} - \beta(\mathrm{tr}\,\mathbf{E})\mathbf{1} - 2\mu \overset{\circ}{\mathbf{E}},$$

a superposed ring denoting the trace-free part, i.e., $\overset{\circ}{\mathbf{E}} = \mathbf{E} - \frac{1}{3}(\mathrm{tr}\mathbf{E})\mathbf{1}$. The quantities α, β_0, β, μ_0, μ are supposed to be real constants. Indeed, β_0 and μ_0 can be viewed as the instantaneous bulk and shear elasticities and then we let $\beta_0 > 0$, $\mu_0 > 0$. Moreover, by stability requirements we let $\alpha > 0$. Thermodynamic requirements [100] imply that $\beta, \mu > 0$. A trivial integration yields

$$\boldsymbol{\xi}(t) = -\beta\left[\int_0^\infty \exp(-\alpha s)\,\mathrm{tr}\,\mathbf{E}(t-s)\,ds\right]\mathbf{1} - 2\mu\int_0^\infty \exp(-\alpha s)\overset{\circ}{\mathbf{E}}(t-s)\,ds,$$

thus showing that this case corresponds to a Boltzmann function of exponential form. The instance when the Boltzmann function is a combination of N exponentials is modelled by an N-tuple of symmetric tensor functions $\boldsymbol{\xi}_i$, $i = 1, \cdots, N$, governed by

$$\dot{\boldsymbol{\xi}}_i = -\alpha_i\boldsymbol{\xi}_i - \beta_i(\mathrm{tr}\mathbf{E})\mathbf{1} - 2\mu_i \overset{\circ}{\mathbf{E}}$$

where α_i, β_i, $\mu_i \in \mathbb{R}^{++}$. Of course, we let

$$\boldsymbol{\xi} = \sum_{i=1}^{N} \boldsymbol{\xi}_i.$$

Though our analysis is confined to linear viscoelasticity,[11] it is worth glancing at finite viscoelasticity and then deriving the linear model as a particular case. To get a simple description of finite viscoelasticity, we start from a general constitutive assumption and then account for material objectivity and fading memory requirements.

Constitutive assumption of finite viscoelasticity. *The stress is given by a functional of the history of the deformation gradient.*

As usual, we denote separately the dependence on the present value and the (past) history and write

$$\mathbf{T}(t) = \hat{\mathbf{T}}(\mathbf{F}(t), \mathbf{F}^t). \tag{3.1.4}$$

The constitutive functional (3.1.4) is required to satisfy the following principle.

Principle of material objectivity. *The properties of a material should appear the same to all observers.*

Direct consequences of this principle follow by considering a one-parameter family of observers to which we associate a time-dependent rotation tensor function $\mathbf{Q}(\tau) : (-\infty, t] \to \text{Orth}^+$ relative to the given observer. To these observers, the deformation gradient $\mathbf{F}(\tau)$ appears as $\mathbf{Q}(\tau)\,\mathbf{F}(\tau)$. Then by the tensor character of $\mathbf{T}$ we have

$$\mathbf{Q}(t)\,\mathbf{T}(t)\,\mathbf{Q}^T(t) = \hat{\mathbf{T}}(\mathbf{Q}(t)\,\mathbf{F}(t), (\mathbf{Q}\,\mathbf{F})^t).$$

By the Polar Decomposition Theorem we can write

$$\mathbf{Q}(t)\,\mathbf{T}(t)\,\mathbf{Q}^T(t) = \hat{\mathbf{T}}(\mathbf{Q}(t)\,\mathbf{R}(t)\,\mathbf{U}(t), (\mathbf{Q}\,\mathbf{R}\,\mathbf{U})^t).$$

The choice $\mathbf{Q}(\tau) = \mathbf{R}^T(\tau)$, $\tau \in (-\infty, t]$, yields

$$\mathbf{T}(t) = \mathbf{R}(t)\,\hat{\mathbf{T}}(\mathbf{U}(t), \mathbf{U}^t)\,\mathbf{R}^T(t) \tag{3.1.5}$$

whereby the stress depends on the present value of $\mathbf{R}$ but is independent of its history. In other words, apart from the explicit dependence on $\mathbf{R}$ shown by (3.1.5), because of the material objectivity the stress depends on $\mathbf{F}(t)$ and

[11] Mathematical problems in nonlinear viscoelasticity are investigated in a book [112] concerning wave propagation, local and global existence, applications of semigroup theory, and steady flows.

$\mathbf{F}^t$ through $\mathbf{U}(t)$ and $\mathbf{U}^t$ only. Letting $\bar{\mathbf{T}}(t) = \mathbf{R}^T(t)\,\mathbf{T}(t)\,\mathbf{R}(t)$ we can say that $\bar{\mathbf{T}}(t)$ depends on $\mathbf{F}(t)$ and $\mathbf{F}^t$ through $\mathbf{C}(t)$ and $\mathbf{C}^t$.

Among the possible representations of the functional (3.1.4), it is worth considering that of Coleman and Noll [26], [27] in terms of the relative tensor $\bar{\mathbf{C}}_t = \mathbf{R}^T(t)\,\mathbf{C}_t(\tau)\,\mathbf{R}(t)$, namely the invariant relative right Cauchy-Green tensor. The functional for $\bar{\mathbf{T}}$ is taken in the form

$$\bar{\mathbf{T}}(t) = \mathbf{T}_e(\mathbf{C}(t)) + \hat{\mathbf{T}}(\mathbf{C}(t), \bar{\mathbf{C}}_t - \mathbf{1}), \tag{3.1.6}$$

subject to the condition that

$$\hat{\mathbf{T}}(\mathbf{C}(t), \mathbf{0}) = 0 \tag{3.1.7}$$

whereby $\mathbf{T}_e$ is the equilibrium response function which gives the value of $\bar{\mathbf{T}}$ when the material has been held at rest up to time t.

The principle of fading memory is made operative by requiring that *the functional* $\hat{\mathbf{T}}$ *is Fréchet differentiable in* $\mathcal{H}$*, uniformly in the tensor parameter* $\mathbf{C}(t)$*, and* $\mathbf{T}_e$ *is continuously differentiable.*

By use of the Riesz representation theorem we can write

$$d\hat{\mathbf{T}}(\mathbf{C}(t), \mathbf{0}|\bar{\mathbf{C}}_t - \mathbf{1}) = \int_0^\infty \mathbf{\Gamma}(s; \mathbf{C}(t))[\bar{\mathbf{C}}_t(t-s) - \mathbf{1}]\,ds$$

where $\mathbf{\Gamma}(\cdot, \mathbf{C}) : \mathbb{R}^+ \to \mathrm{Lin(Sym)}$. The theory based on the representation

$$\bar{\mathbf{T}}(t) = \mathbf{T}_e(\mathbf{C}(t)) + \int_0^\infty \mathbf{\Gamma}(s; \mathbf{C}(t))[\bar{\mathbf{C}}_t(t-s) - \mathbf{1}]\,ds \tag{3.1.8}$$

is called *finite linear viscoelasticity.* The passage to linear (infinitesimal) viscoelasticity is made as follows. Letting

$$\varepsilon = \sup|\mathbf{F}(\tau) - \mathbf{1}|, \qquad \tau \in (-\infty, t],$$

we have

$$\bar{\mathbf{C}}_t(t-s) - \mathbf{1} = 2[\mathbf{E}(t-s) - \mathbf{E}(t)] + o(\varepsilon)$$

and

$$\bar{\mathbf{T}}(t) - \mathbf{T}_e(\mathbf{C}(t)) = \mathbf{\Phi}_0\,\mathbf{E}(t) + \int_0^\infty \dot{\mathbf{\Phi}}(s)\,\mathbf{E}(t-s)\,ds + o(\varepsilon)$$

where

$$\mathbf{\Phi}(s) = -2\int_s^\infty \mathbf{\Gamma}(\xi)\,d\xi, \qquad \mathbf{\Phi}_0 = \mathbf{\Phi}(0)$$

and, by (3.1.7),

$$\lim_{s\to\infty} \mathbf{\Phi}(s) = 0.$$

Similarly we have

$$\mathbf{T}_e(\mathbf{C}(t)) = \mathbf{T}_r + \mathbf{\Lambda}\,\mathbf{E}(t) + o(\varepsilon)$$

where $\mathbf{\Lambda} \in \mathrm{Lin(Sym)}$ and $\mathbf{T}_r$ is the residual stress, namely the stress the material would sustain if it had been held in the reference placement up to the present time. It is apparent that the constitutive equation (3.1.2) of linear (infinitesimal) viscoelasticity follows by disregarding the $o(\varepsilon)$ terms and letting $\mathbf{T}_r = 0$.

3.2. Thermodynamic restrictions for the linear viscoelastic solid. Consistent with the model of viscoelastic solid, whereby the power $\mathbf{T} \cdot \mathbf{D}$ is approximated with $\mathbf{T} \cdot \dot{\mathbf{E}}$, we restate the second law through the work inequality by saying that, for each $\varepsilon > 0$, there exists $\nu_\varepsilon > 0$ such that[12]

$$\int_0^d \boldsymbol{\mathcal{T}}(\mathbf{E}^t) \cdot \dot{\mathbf{E}}(t)\, dt > -\varepsilon \tag{3.2.1}$$

for any ν_ε-approximate cycle. Upon substitution of the constitutive functional (3.1.2) it follows from (3.2.1) that

$$\int_0^d \dot{\mathbf{E}}(t) \cdot \mathbf{G}_0\, \mathbf{E}(t)\, dt + \int_0^d \dot{\mathbf{E}}(t) \cdot \int_0^\infty \mathbf{G}'(s)\mathbf{E}^t(s)\, ds > -\varepsilon. \tag{3.2.2}$$

Of course, in the case of cycles, the inequality (3.2.2) is replaced by

$$\int_0^d \dot{\mathbf{E}}(t) \cdot \mathbf{G}_0\, \mathbf{E}(t)\, dt + \int_0^d \dot{\mathbf{E}}(t) \cdot \int_0^\infty \mathbf{G}'(s)\mathbf{E}^t(s)\, ds \geq 0. \tag{3.2.3}$$

The second law of thermodynamics requires that (3.2.3) hold for any cycle and that equality hold if and only if $\mathbf{E}^t$ is a constant history.

The functional (3.1.2) is compatible with thermodynamics if and only if (3.2.2) is satisfied for approximate cycles. As a particular case, compatibility with thermodynamics means that (3.2.3) holds for cycles. Then we begin by deriving thermodynamic restrictions on $\boldsymbol{\mathcal{T}}$.

Besides being important for easy correlation with experiments [5], time-harmonic variation proves to be especially suited to the derivation of thermodynamic restrictions on the relaxation function [52]–[54]. Accordingly we consider oscillatory strain tensor evolutions of the form

$$\tilde{\mathbf{E}}(t) = \mathbf{E}_1 \cos \omega t + \mathbf{E}_2 \sin \omega t, \tag{3.2.4}$$

where $\omega \in \mathbb{R}^{++}$, $\mathbf{E}_1, \mathbf{E}_2 \in \mathrm{Sym}$. The corresponding process $\tilde{P} \in \Pi$ may be expressed as

$$\tilde{P}(t) = -\omega \mathbf{E}_1 \sin \omega t + \omega \mathbf{E}_2 \cos \omega t, \qquad t \in [0, d), \tag{3.2.5}$$

[12]Henceforth d stands for the duration of the process under consideration.

where $d = 2\pi m/\omega$, m being any positive integer. Accordingly, it follows at once that any current state

$$\tilde{\sigma}(t) = (\tilde{\mathbf{E}}(t),\, {}_r\tilde{\mathbf{E}}^t), \tag{3.2.6}$$

and the process $\tilde{P}$, on $[t, t+d)$, constitute a cycle in that

$$\varrho(\tilde{\sigma}(t),\, \tilde{P}) = \tilde{\sigma}(t).$$

Then $\tilde{\sigma}(t), \tilde{P}$ must satisfy the second law in the form (3.2.3).

THEOREM 3.2.1. *The inequality* (3.2.3) *holds for any cycle* $(\tilde{\sigma}(t), \tilde{P})$ *only if the inequality*

$$\begin{aligned}\mathbf{E}_1 \cdot [\mathbf{G}_0^T - \mathbf{G}_0]\,\mathbf{E}_2 - \int_0^\infty [\mathbf{E}_1 \cdot \mathbf{G}'(s)\mathbf{E}_1 + \mathbf{E}_2 \cdot \mathbf{G}'(s)\mathbf{E}_2] \sin \omega s\, ds \\ - \int_0^\infty \mathbf{E}_1 \cdot [\mathbf{G}'(s) - \mathbf{G}'^T(s)]\mathbf{E}_2 \cos \omega s\, ds \geq 0\end{aligned} \tag{3.2.7}$$

holds for every $\omega \in \mathbb{R}^{++}$ *and every* $\mathbf{E}_1, \mathbf{E}_2 \in \mathrm{Sym}$.

Proof. Consider the cycle $(\tilde{\sigma}(t), \tilde{P}) \in \Sigma \times \Pi$. Substitution into the inequality (3.2.3) gives

$$\begin{aligned}&\int_0^d (-\omega \mathbf{E}_1 \sin \omega t + \omega \mathbf{E}_2 \cos \omega t) \cdot \mathbf{G}_0(\mathbf{E}_1 \cos \omega t + \mathbf{E}_2 \sin \omega t)\, dt \\ &+ \int_0^d \{(-\omega \mathbf{E}_1 \sin \omega t + \omega \mathbf{E}_2 \cos \omega t) \\ &\cdot \int_0^\infty \mathbf{G}'(s)[\mathbf{E}_1(\cos \omega t\, \cos \omega s + \sin \omega t\, \sin \omega s) \\ &+ \mathbf{E}_2(\sin \omega t\, \cos \omega s - \cos \omega t\, \sin \omega s)]ds\}dt \geq 0.\end{aligned}$$

Integration with respect to t, with $d = 2\pi m/\omega$, yields (3.2.7). □

In the limiting cases $\omega \to \infty$, $\omega \to 0$ the inequality (3.2.7) yields some simple, important results.

COROLLARY 3.2.1. *The inequality* (3.2.7) *implies the symmetry of* $\mathbf{G}_0$, *i.e.,*

$$\mathbf{G}_0 = \mathbf{G}_0^T. \tag{3.2.8}$$

Proof. By Riemann-Lebesgue's lemma, the limit $\omega \to \infty$ makes the integrals in (3.2.7) vanish. Then the arbitrariness of $\mathbf{E}_1, \mathbf{E}_2 \in \mathrm{Sym}$ implies (3.2.8). □

COROLLARY 3.2.2. *The inequality* (3.2.7) *implies the symmetry of* $\mathbf{G}_\infty$, *i.e.,*

$$\mathbf{G}_\infty = \mathbf{G}_\infty^T. \tag{3.2.9}$$

Proof. By (3.2.8), in the limiting case $\omega \to 0$ (3.2.7) gives

$$\mathbf{E}_1 \cdot [\mathbf{G}_\infty^T - \mathbf{G}_\infty]\mathbf{E}_2 \geq 0.$$

The arbitrariness of $\mathbf{E}_1, \mathbf{E}_2$ leads to (3.2.9). □

By (3.2.8) we have the following result.

COROLLARY 3.2.3. *The inequality* (3.2.7) *implies that*

$$\int_0^\infty [\mathbf{E}_1 \cdot \mathbf{G}'(s)\mathbf{E}_1 + \mathbf{E}_2 \cdot \mathbf{G}'(s)\mathbf{E}_2] \sin\omega s \, ds + \int_0^\infty \mathbf{E}_1 \cdot [\mathbf{G}'(s) - \mathbf{G}'^T(s)]\mathbf{E}_2 \cos\omega s \, ds \leq 0 \tag{3.2.10}$$

for every $\omega \in \mathbb{R}^{++}$ *and every* $\mathbf{E}_1, \mathbf{E}_2 \in \mathrm{Sym}$.

Let $\mathbf{G}'_s$ on $\mathbb{R}^+$ be the half-range Fourier sine transform of $\mathbf{G}'$, namely

$$\mathbf{G}'_s(\omega) = \int_0^\infty \mathbf{G}'(u) \sin\omega u \, du.$$

The next developments will show the vital role played by the the following restriction on $\mathbf{G}'_s$.

COROLLARY 3.2.4. *The inequality* (3.2.7) *implies the negative definiteness of* $\mathbf{G}'_s$ *in* Sym *for every* $\omega \in \mathbb{R}^{++}$.

Proof. Letting $\mathbf{E}_1 = \mathbf{E}_2$ in (3.2.10) we have $\mathbf{G}'_s(\omega) \leq 0, \omega \in \mathbb{R}^+$, in Sym. By the second law (cf. §2.5), the equality sign holds if and only if the history of $\mathbf{E}$ is constant, namely $\omega = 0$. Then, in Sym,

$$\mathbf{G}'_s(\omega) < 0, \qquad \omega \in \mathbb{R}^{++}, \tag{3.2.11}$$

while $\mathbf{G}'_s(0) = 0$ is identically true. □

COROLLARY 3.2.5. *The inequality* (3.2.11) *implies that, in* Sym,

$$\mathbf{G}_0 - \mathbf{G}(s) > 0, \qquad s \in \mathbb{R}^{++}. \tag{3.2.12}$$

Proof. Upon extending the definition of $\mathbf{G}'(s)$ to $\mathbb{R}^-$ by letting $\mathbf{G}'(-s) = -\mathbf{G}'(s)$ we can write the Fourier sine transform as

$$\mathbf{G}'_s(\omega) = \int_0^\infty \sin\omega s \, \mathbf{G}'(s) \, ds$$

and the inversion formula

$$\mathbf{G}'(s) = \frac{2}{\pi} \int_0^\infty \sin\omega s \, \mathbf{G}'_s(\omega) \, d\omega.$$

Hence the obvious integration with respect to s gives

$$\mathbf{G}(s) - \mathbf{G}_0 = \frac{2}{\pi} \int_0^\infty \frac{1 - \cos \omega s}{\omega} \mathbf{G}'_s(\omega) \, d\omega.$$

The inequality (3.2.11) provides the desired result. □

For ease in writing let

$$\mathbf{G}'_0 := \mathbf{G}'(0).$$

COROLLARY 3.2.6. *In* Sym,

$$\mathbf{G}'_0 \leq 0, \tag{3.2.13}$$

$$\mathbf{G}_0 - \mathbf{G}_\infty \geq 0. \tag{3.2.14}$$

Proof. By (3.2.12), applying the definition of derivative with respect to s provides (3.2.13) while the limit $s \to \infty$ gives (3.2.14). □

In essence these results have already appeared in the literature though they were derived through various approaches and at different times. Here we mention the main pertinent references.

The symmetry of the instantaneous elastic modulus (3.2.8) was proved by Coleman [20] to follow from the second law in the form of the Clausius-Duhem inequality. The symmetry of the equilibrium elastic modulus (3.2.9) was obtained by Day [37] via the Clausius inequality. Apart from the inequality being strict, (3.2.11) was first derived by Graffi [66] in the case of isotropic materials by requiring that energy be dissipated in a period of a sinusoidal strain function $\mathbf{E}(t) = \mathbf{E} \sin \omega t$. Accordingly, (3.2.11) may be rightly referred to as Graffi's inequality.

The connection between (3.2.11) and the energy dissipation is emphasized in [93, §13], where the energy dissipated in one period $[0, d]$, $d = 2\pi/\omega$, is shown to be

$$\int_0^d \boldsymbol{T}(\mathbf{E}^t) \cdot \dot{\mathbf{E}}(t) dt = -\pi \mathbf{E} \cdot \mathbf{G}'_s(\omega) \mathbf{E}.$$

Incidentally, that is why $-\mathbf{G}'_s(\omega)$ is often called the *loss modulus.* The inequality (3.2.13) for the initial derivative of the relaxation function was proved first by Bowen and Chen [8], by having recourse to discontinuous histories, in the one-dimensional case via the Clausius-Duhem inequality. The same result was proved in [99] with C^∞ histories in the three-dimensional case. The inequality (3.2.14) traces back to Coleman [19], [20].

Seemingly the inequality (3.2.12) appeared in [53] for the first time but is in a sense related to a previous result by Day [37] (cf. also [131]) who showed that, as a consequence of dissipativity, the relaxation function satisfies the condition

$$\mathbf{G}_0 - \mathbf{G}_\infty \geq \pm[\mathbf{G}(s) - \mathbf{G}_\infty]. \tag{3.2.15}$$

To show the connection, observe that the limit $s \to \infty$ in the expression for $\mathbf{G}_0 - \mathbf{G}(s)$ gives

$$\mathbf{G}_\infty - \mathbf{G}_0 = \frac{2}{\pi}\int_0^\infty \frac{1}{\omega}\,\mathbf{G}'_s(\omega)\,d\omega.$$

Hence

$$\mathbf{G}(s) - \mathbf{G}_\infty = -\frac{2}{\pi}\int_0^\infty \frac{\cos\omega s}{\omega}\,\mathbf{G}'_s(\omega)\,d\omega.$$

Then the obvious inequalities

$$\int_0^\infty \frac{1}{\omega}|\mathbf{E}\cdot\mathbf{G}'_s(\omega)\mathbf{E}|d\omega \geq \int_0^\infty \frac{|\cos\omega s|}{\omega}|\mathbf{E}\cdot\mathbf{G}'_s(\omega)\mathbf{E}|d\omega$$
$$\geq \left|\int_0^\infty \frac{\cos\omega s}{\omega}\,\mathbf{E}\cdot\mathbf{G}'_s(\omega)\mathbf{E}\,d\omega\right|$$

for any $\mathbf{E} \in \mathrm{Sym}$ provide (3.2.15).

As an aside, the fact that (3.2.12) implies (3.2.15) suggests that, in linear viscoelasticity, the dissipativity holds as a consequence of the second law. We show in §3.4 that such is really the case.

3.3. A sufficiency condition for the second law to hold. The set of inequalities derived in the previous section comprises the thermodynamic inequalities already known. What is more, they are also sufficient for the second law to hold in the general form of the Clausius property. To show that this is so, we consider a fading memory space with an influence function k such that

$$k(s+u) \leq \frac{\gamma}{(\beta+u)^{2+\kappa}}\,k(s) \tag{3.3.1}$$

for any $s, u \in \mathbb{R}^+$ with β, γ, κ suitable positive parameters.

DEFINITION 3.3.1. For any history $\mathbf{E}^t \in \mathcal{H}$ we define the periodic history $\mathbf{E}^t_\tau$, of period τ, as

$$\mathbf{E}^t_\tau(s) = \mathbf{E}^t(s - m\tau), \qquad m\tau \leq s(m+1)\tau, \qquad m = 0, 1, 2, \cdots. \tag{3.3.2}$$

The following theorem ascribes to the set $\mathcal{P}$ of all periodic histories a prominent role.

THEOREM 3.3.1. *The set $\mathcal{P}$ of all periodic histories belonging to $\mathcal{H}$ is dense in $\mathcal{H}$.*

Proof. For any history $\mathbf{E}^t \in \mathcal{H}$ consider the periodic history $\mathbf{E}^t_{\tau_n}$ defined as

$$\mathbf{E}^t_{\tau_n}(s) = \mathbf{E}^t(s - m\tau_n), \quad m\tau_n \leq s < (m+1)\tau_n, \quad m = 0, 1, 2, \cdots, \tag{3.3.3}$$

τ_n being the period; n is any natural number and we let $\tau_n \to \infty$ as $n \to \infty$. The set $\{\mathbf{E}^t_{\tau_n}\}, n = 1, 2, \cdots$, constitutes a sequence of elements of $\mathcal{H}$. Now we prove that $\mathbf{E}^t_{\tau_n}$ tends to $\mathbf{E}^t$, namely

$$\lim_{n\to\infty} \|\mathbf{E}^t_{\tau_n} - \mathbf{E}^t\| = 0. \tag{3.3.4}$$

Since $\mathbf{E}^t_{\tau_n}(s) - \mathbf{E}^t(s) = 0$ as $s \in [0, \tau_n)$, we have

$$\begin{aligned}\|\mathbf{E}^t_{\tau_n} - \mathbf{E}^t\|^2 &= \int_{\tau_n}^{\infty} |\mathbf{E}^t_{\tau_n}(s) - \mathbf{E}^t(s)|^2 k(s)\, ds \\ &\leq 2\int_{\tau_n}^{\infty} [|\mathbf{E}^t_{\tau_n}(s)|^2 + |\mathbf{E}^t(s)|^2] k(s)\, ds.\end{aligned}$$

By the finiteness of $\|\mathbf{E}^t\|$ it follows that

$$\lim_{n\to\infty} \int_{\tau_n}^{\infty} |\mathbf{E}^t(s)|^2 k(s)\, ds = 0.$$

Meanwhile, on account of (3.3.1) we have

$$\int_{\tau_n}^{\infty} |\mathbf{E}_{\tau_n}|^2 k(s)\, ds = \sum_{r=1}^{\infty} \int_{r\tau_n}^{(r+1)\tau_n} |\mathbf{E}^t(s - r\tau_n)|^2 k(s)\, ds \leq E_M^2 \frac{\gamma}{\tau_n^{1+\kappa}} \sum_{r=1}^{\infty} \frac{1}{r^{1+\kappa}}$$

where $E_M =\!\!\to \sup |\mathbf{E}^t(s)|$ as $s \in \mathbb{R}^+$. Then

$$\lim_{n\to\infty} \int_{\tau_n}^{\infty} |\mathbf{E}^t_{\tau_n}(s)|^2 k(s)\, ds = 0.$$

Accordingly (3.3.3) holds and hence $\mathcal{P}$ is dense in $\mathcal{H}$. □

A second, preliminary result concerns the continuity of the work functional relative to the state. To be precise, let $\mathbf{E}^{t+u}, u \in [0, d)$, be the one-parameter family of histories induced by the history $\mathbf{E}^t$ and the process $\dot{\mathbf{E}}_d$ via

$$\mathbf{E}^{t+u}(s) = \begin{cases} \mathbf{E}^t(s-u), & s \in [u, \infty), \\ \mathbf{E}^t(0) + \int_0^{u-s} \dot{\mathbf{E}}_d(\xi)\, d\xi, & s \in [0, u). \end{cases}$$

Then let $w(\mathbf{E}^t, \dot{\mathbf{E}}_d) : \mathcal{H} \times P \to \mathbb{R}$ be the work performed by the stress along the family of histories $\mathbf{E}^{t+u}$, namely

$$w(\mathbf{E}^t, \dot{\mathbf{E}}_d) = \int_0^d \boldsymbol{T}(\mathbf{E}^{t+u}) \cdot \dot{\mathbf{E}}_d(u)\, du.$$

The desired continuity of w is provided by the following theorem.

THEOREM 3.3.2. *The functional* $w(\cdot,\dot{\mathbf{E}}_d):\mathcal{H}\to\mathbb{R}$ *is continuous in* $\mathcal{H}$ *for each bounded process* $\dot{\mathbf{E}}_d$, $|\dot{\mathbf{E}}_d|\le M<\infty$.

Proof. For any two histories $\mathbf{E}_1^{t_1},\mathbf{E}_2^{t_2}\in\mathcal{H}$ we have

$$|w(\mathbf{E}_1^{t_1},\dot{\mathbf{E}}_d)-w(\mathbf{E}_2^{t_2},\dot{\mathbf{E}}_d)|\le\int_0^d|\boldsymbol{T}(\mathbf{E}_1^{t_1+u})-\boldsymbol{T}(\mathbf{E}_2^{t_2+u})||\dot{\mathbf{E}}_d(u)|\,du.$$

By the continuity of $\boldsymbol{T}$, for each $\delta>0$ there exists $\lambda_\delta>0$ such that

$$|\boldsymbol{T}(\mathbf{E}_1^{t_1})-\boldsymbol{T}(\mathbf{E}_2^{t_2})|<\delta\quad\text{whenever }\|\mathbf{E}_1^{t_1+u}-\mathbf{E}_2^{t_2+u}\|<\lambda_\delta.$$

By the definition of $\mathbf{E}^{t+u}$, for each $\lambda_\delta>0$ and $u\in[0,d)$, there exists $\mu_d>0$ such that

$$\|\mathbf{E}_1^{t_1+u}-\mathbf{E}_2^{t_2+u}\|<\lambda_\delta\quad\text{whenever }\|\mathbf{E}_1^{t_1}-\mathbf{E}_2^{t_2}\|<\mu_d.$$

Accordingly, for each $\delta>0$ there exists $\mu_d>0$ such that

$$|w(\mathbf{E}_1^{t_1},\dot{\mathbf{E}}_d)-w(\mathbf{E}_2^{t_2},\dot{\mathbf{E}}_d)|<M\,d\,\delta\quad\text{whenever }\|\mathbf{E}_1^{t_1}-\mathbf{E}_2^{t_2}\|<\mu_d.$$

This proves the continuity of $w(\cdot,\dot{\mathbf{E}}_d)$ in $\mathcal{H}$. □

We are now in a position to prove the sufficiency property.

THEOREM 3.3.3. *If* $\mathbf{G}$ *satisfies* (3.2.8) *and* (3.2.10) *for any* $\omega\in\mathbb{R}^+$ *and* $\mathbf{E}_1,\mathbf{E}_2\in\mathrm{Sym}$ *then the Clausius property holds.*

Proof. As a preliminary step we prove that periodic histories meet (3.2.3).

Consider the periodic function $\mathbf{E}_{\tau_n}:\mathbb{R}\to\mathrm{Sym}$ with period τ_n. The periodic history $\mathbf{E}_{\tau_n}^t\in\mathcal{P}$ can be expressed through its Fourier series as

$$\mathbf{E}_{\tau_n}^t(s)=\sum_{k=0}^{\infty}\mathbf{A}_k\cos k\omega_n(t-s)+\mathbf{B}_k\sin k\omega_n(t-s),\quad\omega_n=\frac{2\pi}{\tau_n},\quad\mathbf{A}_k,\mathbf{B}_k\in\mathrm{Sym};$$

the convergence of the series is meant in the sense of the norm of $\mathcal{H}$. As u varies from 0 to τ_n, the family of histories $\mathbf{E}_{\tau_n}^{t+u}$ constitutes a cycle. The work performed along this cycle is given by

$$\begin{aligned}w=&\int_0^{\tau_n}\dot{\mathbf{E}}_{\tau_n}(t)\cdot\mathbf{G}_0\mathbf{E}_{\tau_n}(t)\,dt\\&+\int_0^{\tau_n}\int_0^{\infty}\sum_{k=1}^{\infty}\sum_{h=0}^{\infty}k\omega_n[-\mathbf{A}_k\sin k\omega_n t+\mathbf{B}_k\cos k\omega_n t]\cdot\mathbf{G}'(s)\\&\qquad[\mathbf{A}_h\cos h\omega_n(t-s)+\mathbf{B}_h\sin h\omega_n(t-s)]ds\,dt.\end{aligned}$$

The first integral vanishes because of the symmetry condition (3.2.8) and the cycle property $\mathbf{E}_{\tau_n}(\tau_n)=\mathbf{E}_{\tau_n}(0)$. Term by term integration of the double series shows that the only nonzero terms are those with $h = k$. Then we are left with

$$w = \sum_{k=1}^{\infty} \int_0^{\infty} \left[\int_0^{\tau_n} k\omega_n \sin^2 k\omega_n t \, dt \right]$$

$$\int_0^{\infty} [-\mathbf{A}_k \cdot \mathbf{G}'(s)\mathbf{A}_k \sin k\omega_n s - \mathbf{A}_k \cdot \mathbf{G}'(s)\mathbf{B}_k \cos k\omega_n s] ds$$

$$+ \sum_{k=1}^{\infty} \int_0^{\infty} \left[\int_0^{\tau_n} k\omega_n \cos^2 k\omega_n t \, dt \right]$$

$$\int_0^{i} nfty[\mathbf{B}_k \cdot \mathbf{G}'(s)\mathbf{A}_k \cos k\omega_n s - \mathbf{B}_k \cdot \mathbf{G}'(s)\mathbf{B}_k \sin k\omega_n s] ds$$

whence we have

$$w = -\pi \sum_{k=1}^{\infty} k \left\{ \int_0^{\infty} [\mathbf{A}_k \cdot \mathbf{G}'(s)\mathbf{A}_k + \mathbf{B}_k \cdot \mathbf{G}'(s)\mathbf{B}_k] \sin k\omega_n s \, ds \right.$$
$$\left. + \int_0^{\infty} \mathbf{A}_k \cdot [\mathbf{G}'(s) - \mathbf{G}'^T(s)]\mathbf{B}_k \cos k\omega_n s \, ds \right\}.$$

The fact that (3.2.10) holds for any $\omega \in \mathbb{R}^+$ and $\mathbf{E}_1, \mathbf{E}_2 \in \text{Sym}$ yields $w = 0$. This proves that periodic histories satisfy (3.2.3).

Now we can prove that (3.2.8) and (3.2.10) ensure the validity of the Clausius property in the form (3.2.1) or (3.2.2).

Let $\mathbf{E}^t \in \mathcal{H}$ be a given history. For every $\lambda > 0$, there exist histories $\bar{\mathbf{E}}^t \in \mathcal{H}$ such that $\bar{\mathbf{E}}^t \in \mathcal{O}_\lambda(\mathbf{E}^t)$ and $\bar{\mathbf{E}}^{t-\tau} = \mathbf{E}^t$, with τ possibly dependent on λ. For any $\bar{\mathbf{E}}^t$ define the periodic history $\bar{\mathbf{E}}^t_\tau$, with period τ, as in (3.3.2). Then we have

$$\|\bar{\mathbf{E}}^t_\tau - \bar{\mathbf{E}}^t\| < \alpha \, \lambda \tag{3.3.5}$$

where

$$\alpha = \frac{\sqrt{\gamma}}{\tau^{1+\kappa/2}} \sum_{m=1}^{\infty} \frac{1}{m^{1+\kappa/2}}.$$

To prove (3.3.5) let $\mathbf{E}^t_n, n = 1, 2, \cdots$, be the sequence of histories defined by

$$\mathbf{E}^t_n(s) = \begin{cases} \bar{\mathbf{E}}^t_\tau(s), & s \in [0, n\tau), \\ \bar{\mathbf{E}}^t(s - n\tau), & s \in [n\tau, \infty). \end{cases}$$

By definition $\mathbf{E}^t_1 = \bar{\mathbf{E}}^t$. Moreover, since $\bar{\mathbf{E}}^t_\tau = \lim \mathbf{E}^t_n$ as $n \to \infty$, we can write

$$\|\bar{\mathbf{E}}^t_\tau - \bar{\mathbf{E}}^t\| = \lim_{n\to\infty} \|\mathbf{E}^t_n - \bar{\mathbf{E}}^t\| \le \lim_{n\to\infty} \sum_{m=1}^{\infty} \|\mathbf{E}^t_{m+1} - \mathbf{E}^t_m\|.$$

For any integer m we have

$$\int_0^\infty |\mathbf{E}^t_{m+1}(s) - \mathbf{E}^t_m(s)|^2\, k(s)\, ds = \int_0^\infty |\bar{\mathbf{E}}^t(s) - \mathbf{E}^t(s)|^2 k(s+m\tau)\, ds$$

and, in view of (3.3.1),

$$\int_0^\infty |\bar{\mathbf{E}}^t(s) - \mathbf{E}^t(s)|^2 k(s+m\tau)\, ds \leq \frac{\gamma}{(\beta+m\tau)^{2+\kappa}} \int_0^\infty |\bar{\mathbf{E}}^t(s) - \mathbf{E}^t(s)|^2 k(s)\, ds,$$

the last integral being smaller than λ^2 because $\bar{\mathbf{E}}^t \in \mathcal{O}(\mathbf{E}^t)$. Hence we obtain

$$\|\bar{\mathbf{E}}^t_\tau - \bar{\mathbf{E}}^t\| \leq \lim_{n\to\infty} \sum_{m=1}^{n} \frac{\sqrt{\gamma}}{(m\tau)^{1+\kappa/2}}\, \lambda$$

whence (3.3.5).

By (3.3.5) we have

$$\|\bar{\mathbf{E}}^t_\tau - \mathbf{E}^t\| \leq \|\bar{\mathbf{E}}^t_\tau - \bar{\mathbf{E}}^t\| + \|\bar{\mathbf{E}}^t - \mathbf{E}^t\| \leq (\alpha+1)\lambda,$$

which makes it always possible to get $\bar{\mathbf{E}}^t_\tau \in \mathcal{O}_\lambda(\mathbf{E}^t)$. Then, because of the continuity of $w(\cdot, \dot{\mathbf{E}}_\tau)$, for every $\delta > 0$ there exists $\lambda > 0$ such that

$$|w(\mathbf{E}^t, \dot{\mathbf{E}}_\tau) - w(\bar{\mathbf{E}}^t_\tau, \dot{\mathbf{E}}_\tau)| < \delta$$

whenever $\bar{\mathbf{E}}^t_\tau \in \mathcal{O}_\lambda(\mathbf{E}^t)$. Meanwhile, letting $\dot{\mathbf{E}}_\tau = \dot{\bar{\mathbf{E}}}^t_\tau$ in the period $[0, \tau]$, we find $w(\bar{\mathbf{E}}^t_\tau, \dot{\mathbf{E}}_\tau)$ is the work performed along a cycle and then

$$w(\bar{\mathbf{E}}^t_\tau, \dot{\mathbf{E}}_\tau) \geq 0.$$

Hence

$$w(\mathbf{E}^t, \dot{\mathbf{E}}_\tau) \geq -\delta$$

and the proof is complete. □

The meaning of this theorem is remarkable. It shows that (3.2.8) and (3.2.10) are both necessary and sufficient for the validity of the second law in the form of the Clausius property. Since both (3.2.8) and (3.2.10) arise from (3.2.7) which reflects a property occurring with time-harmonic strain functions, Theorem 3.3.3 enables us to say that, in order to test for compatibility with thermodynamics, it is enough to examine what happens in time-harmonic processes.

Remark 3.3.1. We have seen that, in spite of (3.2.8) and (3.2.10) being necessary and sufficient, the relaxation function $\mathbf{G}$ need not be symmetric.

Yet, it is evident how (3.2.10) simplifies if we know from the outset that $\mathbf{G}'$ is symmetric.

3.4. Further properties of the relaxation function and dissipativity. Now we show that further inequalities for the relaxation function hold as a consequence of the second law. Indeed, our starting point consists merely of the constitutive inequality (3.1.1) and the thermodynamic inequality[13]

$$\mathbf{G}'_s(\omega) \leq 0, \qquad \omega \in \mathbb{R}^+. \tag{3.4.1}$$

Next we examine what follows if the strict inequality (3.2.11) is involved instead. Of course, a constant relaxation function is not allowed within linear viscoelasticity and then equality in (3.4.1) cannot hold for every $\omega \in \mathbb{R}^+$.

Consider functions g which vanish identically on $\mathbb{R}^-$. Letting $g \in L^1(\mathbb{R}^+)$, we denote by g_L its Laplace transform, namely

$$g_L(p) = \int_0^\infty \exp(-ps)\, g(s)\, ds$$

where p is a complex number with $\Re p \geq 0$. Similarly, we denote by g_F its Fourier transform, namely

$$g_F(\omega) = \int_0^\infty \exp(-i\omega s)\, g(s)\, ds, \qquad \omega \in \mathbb{R}.$$

Of course, letting the subscript c denote the half-range Fourier cosine transform, we have

$$g_F = g_c - i\, g_s.$$

If, instead, $g \in L^1(\mathbb{R})$ then we set

$$g_F(\omega) = \int_{-\infty}^\infty \exp(-i\omega s)\, g(s)\, ds.$$

Two properties hold depending on whether g is even or odd:

$$g(-s) = g(s),\ s \in \mathbb{R}^{++} \ \rightarrow\ g_F = g_C = 2g_c,$$

$$g(-s) = -g(s),\ s \in \mathbb{R}^{++} \ \rightarrow\ g_F = -ig_S = -2ig_s.$$

Let $\breve{\mathbf{G}}(s) = \mathbf{G}(s) - \mathbf{G}_\infty$. An integration by parts gives

$$\int_0^\infty \mathbf{G}'(s) \sin \omega s\, ds = -\omega \int_0^\infty \breve{\mathbf{G}}(s) \cos \omega s\, ds.$$

[13]This is usually considered in place of (3.2.11) when no distinction is made between reversible and irreversible processes.

This proves the following lemma.

LEMMA 3.4.1. *If* $\mathbf{G}' \in L^1(\mathbb{R}^+)$ *then, for any nonzero* ω,

$$\breve{\mathbf{G}}_c(\omega) = -\frac{1}{\omega}\mathbf{G}'_s(\omega).$$

The following theorems yield two new inequalities for $\mathbf{G}'$.

THEOREM 3.4.1. *If the relaxation function* $\mathbf{G}$ *satisfies* (3.1.1) *and* (3.4.1) *then*

$$\int_0^\infty \mathbf{G}'(s)\exp(-\alpha s)\sin\omega s\, ds < 0, \qquad \alpha \in \mathbb{R}^{++}, \qquad \omega \in \mathbb{R}^{++}. \tag{3.4.2}$$

Proof. By Plancherel's theorem,[14] for any pair of functions $g, h \in L^1 \cap L^2(\mathbb{R})$ the inequality

$$\int_{-\infty}^\infty g(s)\, h^*(s)\, ds = \frac{1}{2\pi}\int_{-\infty}^\infty g_F(\tau)\, h_F^*(\tau)\, d\tau \tag{3.4.3}$$

holds, the symbol $*$ denoting the complex conjugate. Letting $\alpha \in \mathbb{R}^{++}$, $\omega \in \mathbb{R}^+$, we choose g and h as

$$g(s) = \begin{cases} -\mathbf{G}'(-s), & s < 0, \\ \mathbf{G}'(s), & s \geq 0, \end{cases}$$

and

$$h(s) = \begin{cases} 0, & s < 0, \\ \exp[-(\alpha + i\omega)s], & s \geq 0. \end{cases}$$

Obviously both g and h are in $L^1 \cap L^2(\mathbb{R})$. Then by (3.4.3) we have

$$g_L(p) = \frac{1}{2\pi}\int_{-\infty}^\infty g_F(\tau)\,\frac{p + i\tau}{p^2 + \tau^2}\, d\tau$$

while

$$g_F(\tau) = -2i\mathbf{G}'_s(\tau).$$

Letting $p = \alpha + i\omega$ and observing that $\mathbf{G}'_s(\tau)$ is odd (and $\tau\mathbf{G}'_s(\tau)$ is even) in $\mathbb{R}$, we have

$$g_L(p) = \frac{2}{\pi}\int_0^\infty \frac{\tau\mathbf{G}'_s(\tau)}{\alpha^2 - \omega^2 + \tau^2 + 2i\alpha\omega}\, d\tau$$

whence

$$\int_0^\infty \mathbf{G}'(s)\exp(-\alpha s)\sin\omega s\, ds = \frac{4\alpha\omega}{\pi}\int_0^\infty \frac{\tau\mathbf{G}'_s(\tau)}{(\alpha^2 + \tau^2 - \omega^2)^2 + 4\alpha^2\omega^2}\, d\tau.$$

---

[14]Actually, by a consequence of Plancherel's theorem; cf. [6], Thm. 52.

Then (3.4.1) implies (3.4.2).[15] □

THEOREM 3.4.2. *If the relaxation function* $\mathbf{G}$ *satisfies* (3.1.1) *and* (3.2.11) then

$$\mathbf{G}_0 + \int_0^\infty \mathbf{G}'(s)\exp(-\alpha s)\,ds > \mathbf{G}_\infty > 0, \qquad \alpha \in \mathbb{R}^{++}. \tag{3.4.4}$$

Proof. Consider the left-hand side of (3.4.4). An integration by parts shows that

$$\mathbf{G}_0 + \int_0^\infty \mathbf{G}'(s)\exp(-\alpha s)\,ds = \mathbf{G}_\infty + \alpha \int_0^\infty \breve{\mathbf{G}}(s)\exp(-\alpha s)\,ds.$$

By Fourier's integral theorem

$$\breve{\mathbf{G}}(s) = \frac{2}{\pi}\int_0^\infty \breve{\mathbf{G}}_c(\omega)\cos\omega s\,d\omega.$$

Substitution and integration with respect to s yields

$$\mathbf{G}_0 + \int_0^\infty \mathbf{G}'(s)\exp(-\alpha s)\,ds = \mathbf{G}_\infty + \frac{2\alpha^2}{\pi}\int_0^\infty \frac{\breve{\mathbf{G}}_c(\omega)}{\alpha^2+\omega^2}\,d\omega.$$

The observation that $\breve{\mathbf{G}}_c(\omega) = -\mathbf{G}'_s(\omega)/\omega$, along with (3.1.1) and (3.2.11), yields (3.4.4). □

Now we prove the further result that, in linear viscoelasticity, the dissipativity condition is a consequence of the second law of thermodynamics for all histories in $H^1(\mathbb{R}^+)$.

THEOREM 3.4.3. *If the relaxation function* $\mathbf{G}$ *is symmetric and satisfies* (3.1.1) *and* (3.4.1) *then, for any* $\mathbf{E}^t \in H^1(\mathbb{R}^+)$,

$$w(\mathbf{E}^t) := \int_{-\infty}^t \boldsymbol{T}(\mathbf{E}^\tau)\cdot\dot{\mathbf{E}}(\tau)\,d\tau \geq 0. \tag{3.4.5}$$

Proof. Given any history $\mathbf{E}^t \in H^1$ we extend the function $\mathbf{E}$ to $\mathbb{R}$ through the static continuation

$$\varepsilon(\tau) = \begin{cases} \mathbf{E}(\tau), & \tau \leq t, \\ \mathbf{E}(t), & \tau > t. \end{cases}$$

Moreover, we extend $\breve{\mathbf{G}}$ to $\mathbb{R}$ by letting

$$\bar{\mathbf{G}}(s) := \breve{\mathbf{G}}(|s|), \qquad s \in \mathbb{R}.$$

[15] The proof does not apply if $\alpha = 0$ because in that case $h \notin L^1(\mathbb{R}) \cap L^2(\mathbb{R})$.

Then by the symmetry of $\mathbf{G}$ we have

$$w(\mathbf{E}^t) = \frac{1}{2}\mathbf{E}(t)\cdot\mathbf{G}_\infty\mathbf{E}(t) + \frac{1}{2}\int_{-\infty}^{\infty}\dot{\varepsilon}(u)\cdot\int_{-\infty}^{\infty}\bar{\mathbf{G}}(u-s)\dot{\varepsilon}(s)\,ds\,du.$$

Hence, by Plancherel's theorem,

$$w(\mathbf{E}^t) = \frac{1}{2}\mathbf{E}(t)\cdot\mathbf{G}_\infty\mathbf{E}(t) + \frac{1}{4\pi}\int_{-\infty}^{\infty}\dot{\varepsilon}_F^*(\omega)\cdot\bar{\mathbf{G}}_F(\omega)\dot{\varepsilon}_F(\omega)\,d\omega;$$

$\dot{\varepsilon}_F(\omega)$ is parameterized by the time t in that

$$\dot{\varepsilon}_F(\omega) = \int_{-\infty}^{t}\exp(-i\omega\tau)\dot{\mathbf{E}}(\tau)\,d\tau.$$

Then we have

$$w(\mathbf{E}^t) = \frac{1}{2}\mathbf{E}(t)\cdot\mathbf{G}_\infty\mathbf{E}(t) + \frac{1}{2\pi}\int_0^{\infty}[\dot{\varepsilon}_c(\omega)\cdot\bar{\mathbf{G}}_c(\omega)\,\dot{\varepsilon}_c(\omega) + \dot{\varepsilon}_s(\omega)\cdot\bar{\mathbf{G}}_c(\omega)\,\dot{\varepsilon}_s(\omega)]d\omega.$$

Because $\bar{\mathbf{G}}_c(\omega) = -\mathbf{G}'_s(\omega)/\omega$, (3.1.1) and (3.4.1) imply (3.4.5). □

Now consider the strict inequality (3.2.11) along with (3.1.1). While (3.4.2) still holds for any $\alpha \in \mathbb{R}^{++}$, it follows at once by (3.2.11) that it holds also for $\alpha = 0$. Then we can replace (3.4.2) with the stronger result

$$\text{(3.4.6)} \qquad \int_0^{\infty}\mathbf{G}'(s)\exp(-\alpha s)\sin\omega s\,ds < 0, \qquad \alpha \in \mathbb{R}^+, \qquad \omega \in \mathbb{R}^{++}.$$

Hence, by (3.4.4) it follows that $\mathbf{G}_0 + \mathbf{G}'_L(p)$ does not vanish in Sym for any complex $p \in \mathbb{C}^+$. This result allows the extension of existence and uniqueness results of elastostatics to quasi-static viscoelasticity [93].

3.5. Free energy. As also pointed out by Day [40], there are objections to assuming the existence of thermodynamic potentials, like energy, entropy, and free energy, for materials with memory. This is so because quite often thermodynamic potentials turn out to be nonunique. It is then of interest to examine whether, and to what extent, thermodynamic potentials are nonunique in linear viscoelasticity as well.[16] By paralleling a procedure elaborated by Day in [102] the class of free energies has been characterized by making the nonuniqueness apparent. For later convenience, the main results are restated here.

[16]In this regard we have to observe that the isothermal approximation makes the free energy to coincide with the energy.

For a while, let $\boldsymbol{\mathcal{T}}$ be the stress functional but not necessarily that given by (3.1.2). A localised function $\mathbf{f}$ on $\mathbb{R}$ of duration d_f is continuous and piecewise C^2 in $[0, d_f]$ and vanishing elsewhere. Consider processes P in $[0,\ d]$ and, for any $t \in [0, d]$, let $P_t = \dot{\mathbf{E}}_t$. Letting $\mathcal{U}$ be an open, connected subset of Sym, we denote by $\boldsymbol{\mathcal{T}}^\dagger : H \to \text{Sym}$ the tensor function

$$\boldsymbol{\mathcal{T}}^\dagger(\mathbf{A}) = \boldsymbol{\mathcal{T}}(\mathbf{A}^\dagger), \qquad \mathbf{A} \in \mathcal{U}.$$

Let $\overset{v}{\underset{u}{w}}(\mathbf{E}(\cdot))$ be the work exerted by $\mathbf{E}(\cdot)$ between the times $u \leq 0$, $v \geq d$, namely

$$\overset{v}{\underset{u}{w}}(\mathbf{E}(\cdot)) = \int_u^v \boldsymbol{\mathcal{T}}(\mathbf{E}^t) \cdot \dot{\mathbf{E}}(t)dt.$$

For any cycle $(\mathbf{E}^0,\ \dot{\mathbf{E}}_d)$ the Clausius inequality reads

$$\overset{d}{\underset{0}{w}}(\mathbf{E}(\cdot)) \geq 0$$

with $\mathbf{E}^t = \varrho(\mathbf{E}^0,\ \dot{\mathbf{E}}_t)$. We write $\overset{d}{\underset{0}{w}}{}^\dagger(\mathbf{E}(\cdot))$ for $\overset{d}{\underset{0}{w}}$ when $\boldsymbol{\mathcal{T}}$ is replaced by $\boldsymbol{\mathcal{T}}^\dagger$. Under the requirement that $\boldsymbol{\mathcal{T}}$ satisfy

$$\overset{d}{\underset{0}{w}}(\mathbf{E}(\cdot)) \geq 0$$

for every cycle $(\mathbf{E}^0,\ \dot{\mathbf{E}}_d)$, be elastic under retardation,[17] and have fading memory, Day proved the following lemma.

LEMMA 3.5.1. *There exists a smooth function $\phi : \mathcal{U} \to \mathbb{R}$ such that*

$$\boldsymbol{\mathcal{T}}^\dagger(\mathbf{E}) = \frac{\partial \phi}{\partial \mathbf{E}}$$

and, for each process $\dot{\mathbf{E}}_t$,

$$w(\mathbf{E}^0,\ \dot{\mathbf{E}}_t) \geq \phi(\mathbf{E}(t)) - \phi(\mathbf{E}(0)). \tag{3.5.1}$$

The proof is given in [41] and then is omitted.

A free energy functional ψ is characterised by the properties

(F1) $\quad \psi(\mathbf{A}^\dagger) = \phi(\mathbf{A}), \qquad \mathbf{A} \in \mathcal{U}$,

(F2) $\quad \overset{v}{\underset{u}{w}}(\mathbf{E}^t) \geq \psi(\mathbf{E}^v) - \psi(\mathbf{E}^u), \qquad u, v \in \mathbb{R},$

---

[17]Let f be a localised function on $\mathbb{R}$. Then the function f_a defined as $f_a(t) = f(at)$, $t \in [0, d_f]$, $a \in (0, 1)$ is a retardation of f; $\boldsymbol{\mathcal{T}}$ is said elastic under retardation if $\lim w(f_a) = w^\dagger(f)$ as $a \to 0$. The functional (3.1.2) turns out to be elastic under retardation.

(F₃) $\psi(\mathbf{E}^t) \geq \phi(\mathbf{E}(t))$.
The set of such functionals is denoted by $\mathcal{F}$.

It is worth remarking that in theories based on the Clausius-Duhem inequality, as the statement of the second law of thermodynamics, the existence of a free energy functional is assumed at the outset and then the validity of the properties (F_1)–(F_3) is proved.

Let $\mathbf{f} : \mathbb{R} \to \mathcal{U}$ be a localised function. The localised function $\mathbf{g} : \mathbb{R} \to \mathcal{U}$ is a closed continuation of $\mathbf{f}$ at time t if $d_g \geq t$ and $\mathbf{f}^t = \mathbf{g}^t$, $\mathbf{g}(d_g) = \mathbf{f}(t)$. Denote by $\mathcal{C}(\mathbf{f}, t)$ the set of closed continuations of $\mathbf{f}$ at t. The following theorem is proved in [41].

THEOREM 3.5.1. *Let m be the functional on H defined as*

$$m(\mathbf{f}^t) = \sup \left\{ -\overset{d_g}{\underset{t}{w}}(\mathbf{g}) | \mathbf{g} \in \mathcal{C}(\mathbf{f}, t) \right\}.$$

Then $m : H \to \mathbb{R}^+$ and the functional ψ_m on H given by

$$\psi_m(\mathbf{f}^t) = \phi(\mathbf{f}(t)) + m(\mathbf{f}^t)$$

are free energy functionals.

The functional $m(\mathbf{f}^t)$ is called *maximum recoverable work.* It is a prominent result that the set $\mathcal{F}$ has a minimal and a maximal element, the former being ψ_m. This is made precise as follows.

THEOREM 3.5.2. *If $\psi_M : H \to \mathbb{R}$ is defined as*

$$\psi_M(\mathbf{f}^t) = \phi(\mathbf{f}(0)) + \overset{t}{\underset{0}{w}}(\mathbf{f}(\cdot)),$$

then $\psi_M \in \mathcal{F}$. Moreover, for any $\psi \in \mathcal{F}$ and $\mathbf{f}^t \in H$,

$$\psi_m(\mathbf{f}^t) \leq \psi(\mathbf{f}^t) \leq \psi_M(\mathbf{f}^t).$$

Proof. The definition of ψ_M and the inequality (3.5.1) imply that F_3 holds for ψ_M. Now, the obvious property

$$\overset{v}{\underset{u}{w}}(\mathbf{f}(\cdot)) = \overset{v}{\underset{0}{w}}(\mathbf{f}(\cdot)) - \overset{u}{\underset{0}{w}}(\mathbf{f}(\cdot))$$

implies (F_2), in the form of equality, and (F_1). Thus $\psi_M \in \mathcal{F}$.

Let now $\psi \in \mathcal{F}$ and $\mathbf{f}^t \in H$ be given. Consider (F_2) with $u = 0$, $v = t$. Since $\psi(\mathbf{f}^0) = \phi(\mathbf{f}(0))$ then (F_2) and the definition of ψ_M yield

$$\psi(\mathbf{f}^t) \leq \psi_M(\mathbf{f}^t). \tag{3.5.2}$$

For any localised function $\mathbf{g}$, with support $[0, d_g]$, we have

$$\overset{d_g}{\underset{t}{w}}(\mathbf{g}(\cdot)) \geq \psi(\mathbf{g}^{d_g}) - \psi(\mathbf{g}^t).$$

Then, letting $\mathbf{g} \in \mathcal{C}(\mathbf{f}, t)$ we find that

$$\psi(\mathbf{f}^t) \geq \psi(\mathbf{g}^{d_g}) - \overset{d_g}{\underset{t}{w}}(g(\cdot)).$$

Since $\psi(\mathbf{g}^{d_g}) \geq \phi(\mathbf{g}(d_g))$ it follows from (F_3) that

$$\psi(\mathbf{g}^{d_g}) \geq \phi(\mathbf{f}(t))$$

and hence

$$\psi(\mathbf{f}^t) \geq \phi(\mathbf{f}(t)) - \overset{d_g}{\underset{t}{w}}(\mathbf{g}(\cdot)).$$

Taking the supremum of the right-hand side over all $\mathbf{g} \in \mathcal{C}(\mathbf{f}, t)$ and the definition of ψ_m we get

$$\psi(\mathbf{f}^t) \geq \psi^m(\mathbf{f}^t). \tag{3.5.3}$$

The inequalities (3.5.2)–(3.5.3) provide the desired result. □

The set $\mathcal{F}$ is convex in the following sense.

THEOREM 3.5.3. *If $\psi_1, \psi_2 \in \mathcal{F}$ then, for any $\gamma \in [0,1]$, the functional*

$$\psi = \gamma\psi_1 + (1-\gamma)\psi_2$$

is in $\mathcal{F}$.

Proof. Since $\psi_1(\mathbf{A}^\dagger) = \psi_2(\mathbf{A}^\dagger) = \phi(\mathbf{A})$ it follows at once that ψ meets (F_1). Moreover, since ψ_1 and ψ_2 satisfy (F_2), from

$$\psi(\mathbf{f}^v) - \psi(\mathbf{f}^u) = \gamma[\psi_1(\mathbf{f}^v) - \psi_1(\mathbf{f}^u)] + (1-\gamma)[\psi(\mathbf{f}^v) - \psi_2(\mathbf{f}^u)]$$

we have

$$\overset{v}{\underset{u}{w}}(\mathbf{f}(\cdot)) \geq \psi(\mathbf{f}^v) - \psi(\mathbf{f}^u).$$

Finally, $\psi_1(\mathbf{f}^t), \psi_2(\mathbf{f}^t) \geq \phi(\mathbf{f}(t))$ implies that ψ meets (F_3). □

The explicit evaluation of the minimal free energy ψ_m for linear viscoelasticity was performed by Day [38], [41] under the additional assumption that $\mathbf{G}$ is a scalar function, on $\mathbb{R}^+$, given by

$$G(s) = \alpha \exp(-s/\tau) + \beta, \qquad \alpha, \beta \in \mathbb{R}^{++}.$$

It follows that $\phi(\mathbf{A}) = \frac{1}{2}G(\infty)\mathbf{A}^2 = \frac{1}{2}\beta\mathbf{A}^2$ is the equilibrium free energy while

$$\psi_m(\mathbf{E}^t) = \frac{1}{2}\beta\mathbf{E}^2(t) + \frac{1}{2}\alpha\left\{\mathbf{E}(t) - \frac{1}{\tau}\int_0^\infty \exp(-s/\tau)\mathbf{E}^t(s)\,ds\right\}^2.$$

It is then clear that in general the free energy is nonunique. As we show in a moment, uniqueness results by following particular descriptions of the state (via internal variables) or by confining the attention to particular forms of the functional.

To establish a useful connection with other results appearing in the literature, we examine some aspects on the determination of the free energy through the restrictions placed by the Clausius-Duhem inequality. Any continuous stress functional $\boldsymbol{\mathcal{T}}(\mathbf{E}^t)$ and C^1 free energy functional $\psi(\mathbf{E}^t)$, $\mathbf{E}^t \in H$, are compatible with the Clausius-Duhem inequality if and only if [19], [20]

$$\boldsymbol{\mathcal{T}} = \frac{\partial \psi}{\partial \mathbf{E}} \tag{3.5.4}$$

and

$$d\psi(\mathbf{E}(t), {}_r\mathbf{E}^t | {}_r\dot{\mathbf{E}}^t) \leq 0. \tag{3.5.5}$$

Given the stress functional $\boldsymbol{\mathcal{T}}$, we might determine the functional, or the set of functionals, ψ compatible with (3.5.4)–(3.5.5). This problem has been considered in [100, §3.5], with reference to linear viscoelasticity. As a result, it is shown that both

$$\psi_G(\mathbf{E}^t) = \frac{1}{2}\mathbf{E}(t) \cdot \mathbf{G}_\infty \mathbf{E}(t) - \frac{1}{2}\int_0^\infty [\mathbf{E}(t) - \mathbf{E}(t-s)] \cdot \mathbf{G}'(s)[\mathbf{E}(t) - \mathbf{E}(t-s)]\, ds,$$

with $\mathbf{G}'(s) \leq 0$, $\mathbf{G}''(s) \geq 0$, and

$$\psi_D(\mathbf{E}^t) = \frac{1}{2}\mathbf{E}(t) \cdot \mathbf{G}_\infty \mathbf{E}(t) + \frac{1}{2}\left\{(\mathbf{G}_0 - \mathbf{G}_\infty)^{-1/2} \int_0^\infty \mathbf{G}'(s)[\mathbf{E}(t) - \mathbf{E}(t-s)]\, ds\right\}^2,$$

with $\mathbf{G}'(s)$ isotropic and in exponential form, are solutions to (3.5.4)–(3.5.5). Accordingly, nonuniqueness of the free energy also occurs in this framework.

As an aside, the solution ψ_G traces back to Volterra [127]–[129]; its thermodynamic admissibility was proved by Graffi [67]–[69]. The solution ψ_D, instead, generalizes that evaluated by Day.

It is worth observing that there are contexts where the free energy turns out to be unique. One is that of (viscoelastic) materials with internal variables. Consider the model described in §3.1, with $N = 1$. As shown in [100], §14.2, the free energy turns out to be uniquely determined and is given by

$$\psi = \frac{1}{2}\left(\beta_0 + \frac{\beta}{\alpha}\right)(\mathrm{tr}\mathbf{E})^2 + \left(\mu_0 + \frac{\mu}{\alpha}\right)\mathring{\mathbf{E}} \cdot \mathring{\mathbf{E}} + \frac{\beta}{2\alpha}(\mathrm{tr}\mathbf{E} + \mathrm{tr}\boldsymbol{\xi})^2 + \frac{\mu}{\alpha}\left(\mathring{\mathbf{E}} + \frac{\alpha}{2\mu}\mathring{\boldsymbol{\xi}}\right)^2,$$

which is in fact the particular form taken by ψ_D. Meanwhile the procedure shows also that uniqueness may cease to hold if the number of internal variables is greater than that of the physical variables (cf. also [39]). By way of example, this was ascertained in [70], [71] where a Boltzmann function $\mathbf{G}'(s)$ is taken in the form of a linear combination of exponentials, $N = 2$.

Another context where uniqueness of the free energy holds is that of nonlinear viscoelastic materials of single-integral type. Such a model is based on a (one-dimensional) constitutive relation which, in our notation, can be written as

$$\mathcal{T}(E^t) = T(E(t)) + \int_0^\infty \tilde{T}(s, E(t), E(t-s))ds$$

where $T \in C^1(\mathbb{R})$, $\tilde{T} \in C^1(\mathbb{R}^{++}, \mathbb{R}, \mathbb{R})$. The free energy is sought in the single-integral form as

$$\psi(E^t) = \Psi(E(t)) + \int_0^\infty \tilde{\psi}(s, E(t), E(t-s))ds.$$

Gurtin and Hrusa [78] have shown that, upon normalizing the functions $\tilde{T}$ and $\tilde{\psi}$ as

$$\tilde{T}(s, E, E),\ \tilde{\psi}(s, E, E) = 0, \qquad s \in \mathbb{R}^+,$$

for all values of E, there is a free energy of single-integral form if and only if

$$\int_\varepsilon^E \frac{\partial \tilde{T}}{\partial s}(s, \alpha, \varepsilon)\, d\alpha \leq 0, \qquad \varepsilon, E \in \mathbb{R}. \tag{3.5.6}$$

Indeed, the functions Ψ and $\tilde{\psi}$ are given by

$$\Psi(E) = \int_0^E T(\alpha)\, d\alpha,$$

$$\tilde{\psi}(s, E, \varepsilon) = \int_\varepsilon^E \tilde{T}(s, \alpha, \varepsilon)\, d\alpha.$$

In linear viscoelasticity (3.5.6) holds provided that G is convex and in such a case we have

$$\psi(t) = \frac{1}{2} G_\infty E^2(t) - \frac{1}{2} \int_0^\infty G'(s)[E(t) - E(t-s)]^2 ds.$$

Of course, ψ coincides with ψ_G.

By way of example, consider a viscoelastic stress-strain relation where $G'(s) = -\gamma \exp(-\mu s)$, $\gamma, \mu > 0$. The free energy is found to be [78]

$$\psi_1(t) = \frac{1}{2} G_\infty E^2(t) - \frac{1}{2} \int_0^\infty G'(s)[E(t) - E(t-s)]^2\, ds.$$

The single-integral form ψ_1 is merely the free energy ψ_G. Yet we know that ψ_D too, namely

$$\psi_2(t) = \frac{1}{2} G_\infty E^2(t) + \frac{\mu}{2\gamma} \left\{ \int_0^\infty G'(s)[E(t) - E(t-s)]\, ds \right\}^2,$$

provides a free energy for the stress-strain relation. In general, ψ_1 and ψ_2 are substantially different. Indeed, consistent with the minimum property of ψ_D, a direct application of the Cauchy-Schwarz inequality allows us to see that $\psi_1 \geq \psi_2$ [68], [69].

3.6. The linear viscoelastic fluid. *Fluids* are defined as materials for which the symmetry group is the full unimodular group. Then, by starting from a constitutive equation of the form (3.1.8) and letting the current placement be undistorted (namely, $\mathbf{R}(t) = \mathbf{1}$), it can be shown ([25], [43], [46]) that

$$\begin{aligned} \mathbf{T}(t) = -\Pi(\rho(t))\,\mathbf{1} &+ \int_0^\infty \mu'(s;\rho)[\mathbf{C}_t(t-s) - \mathbf{1}]ds \\ &+ \frac{1}{2}\left\{ \int_0^\infty \lambda'(s;\rho)\mathrm{tr}[\mathbf{C}_t(t-s) - \mathbf{1}]ds \right\}\mathbf{1} \end{aligned}$$

where μ' and λ' are scalar relaxation functions and ρ stands for the present value of the mass density. The quantity $\Pi(\rho)$ takes the meaning of equilibrium pressure, namely the pressure corresponding to the fluid being held at rest up to time t. The linear (infinitesimal) theory follows by an analogous procedure to that outlined in §3.1 for the viscoelastic solid. Letting μ', λ' be independent of ρ to the linear order in ε, we have

$$\mathbf{T}(t) = -\Pi(\rho(t))\,\mathbf{1} + 2\int_0^\infty \mu'(s)\,\mathbf{E}_t^t(s)\,ds + \int_0^\infty \lambda'(s)[\mathrm{tr}\mathbf{E}_t^t(s)]ds\,\mathbf{1}.$$

Define

$$\mu(s) = -\int_s^\infty \mu'(\xi)\,d\xi, \qquad \lambda(s) = -\int_s^\infty \lambda'(\xi)\,d\xi$$

and assume that

$$\lim_{s\to\infty} \mu(s),\ \lambda(s) = 0.$$

Since

$$\frac{d}{ds}\mathbf{E}_t^t(s) = -\dot{\mathbf{E}}^t(s), \qquad \mathbf{E}_t^t(0) = 0,$$

an integration by parts gives

$$\mathbf{T}(t) = -\Pi(\rho(t))\,\mathbf{1} + 2\int_0^\infty \mu(s)\dot{\mathbf{E}}^t(s)\,ds + \int_0^\infty \lambda(s)[\mathrm{tr}\dot{\mathbf{E}}^t(s)]\,ds\,\mathbf{1}.$$

Letting, as in §3.1, $\varepsilon = \sup|\mathbf{F}(\tau) - \mathbf{1}|$, $\tau \in (-\infty, t]$, we have

$$\mathbf{D} = \dot{\mathbf{E}} + O(\varepsilon), \qquad \dot{\mathbf{C}} = \dot{\mathbf{E}} + O(\varepsilon).$$

Accordingly, to adhere to a standard notation in fluid dynamics, we write the constitutive equation as

$$\text{(3.6.1)} \quad \mathbf{T}(t) = -\Pi(\rho(t))\,\mathbf{1} + 2\int_0^\infty \mu(s)\mathbf{D}^t(s)\,ds + \int_0^\infty \lambda(s)[\operatorname{tr}\mathbf{D}^t(s)]\,ds\,\mathbf{1}.$$

Quite naturally in this case the fading memory space consists of all histories $\mathbf{D}^t$ with finite k-norm. For later convenience it is worth writing (3.6.1) in terms of tr$\mathbf{D}$ and the trace-free part $\overset{\circ}{\mathbf{D}}$. We have

$$\text{(3.6.2)} \quad \mathbf{T}(t) = -\Pi(\rho(t))\,\mathbf{1} + 2\int_0^\infty \mu(s)\;\overset{\circ}{\mathbf{D}}{}^t(s)\,ds + \int_0^\infty \kappa(s)[\operatorname{tr}\mathbf{D}^t(s)]ds\,\mathbf{1}$$

where $\kappa(s) = \lambda(s) + \frac{2}{3}\mu(s)$ is the bulk relaxation function.

Incidentally, the well-known Navier-Stokes model of viscous fluids follows as a limiting case when $\mu(s)$ and $\lambda(s)$—or $\kappa(s)$—are Dirac's delta functions of the form $\mu\delta(s)$, $\lambda\delta(s)$.

We recall that more general models of fluids with fading memory are often applied. Among them we mention the fluid of second grade [25] where, to within the pressure term, the stress is given by a linear combination of $\mathbf{A}_1, \mathbf{A}_1^2, \mathbf{A}_2$ with $\mathbf{A}_1, \mathbf{A}_2$ the Rivlin-Ericksen tensors

$$\mathbf{A}_1 = \mathbf{D}, \qquad \mathbf{A}_2 = \dot{\mathbf{D}} + \mathbf{D}\,\mathbf{L} + \mathbf{L}^T\,\mathbf{D}.$$

Also we mention the BKZ fluid[18] where $\mathbf{T} + \Pi\,\mathbf{1}$ is a nonlinear functional of the history of $\mathbf{C}_t$. Specifically, the fluid is taken to be incompressible and then $\mathbf{T} + \Pi\,\mathbf{1}$ is expressed as

$$\mathbf{T} + \Pi\,\mathbf{1} = 2\int_0^\infty \left[\frac{\partial W}{\partial I_1}(s, I_1(t-s), I_2(t-s))\mathbf{C}_t^{-1}(t-s) \right.$$
$$\left. - \frac{\partial W}{\partial I_2}(s, I_1(t-s), I_2(t-s))\mathbf{C}_t(t-s)\right] ds$$

where $I_1 = \operatorname{tr}\mathbf{C}_t^{-1}$, $I_2 = \operatorname{tr}\mathbf{C}_t$, and W is a scalar (potential) function. Details on this model and an analysis of three-dimensional motions are exhibited in [112].

---

[18] The model was introduced independently by Kaye [87] and Bernstein, Kearsley, and Zapas [4]. That is why sometimes the model is referred to as K-BKZ fluid.

3.7. Thermodynamic restrictions for the viscoelastic fluid. With reference to the model of viscoelastic fluid characterized by the constitutive functional $\hat{\mathbf{T}}(\rho, \mathbf{D}^t)$ defined by (3.6.1), or (3.6.2), the state σ is the pair of the mass density ρ and the history of $\mathbf{D}$, namely $\sigma(t) = (\rho(t), \mathbf{D}^t)$, while the process is given by

$$P(t) = \mathbf{D}(t), \qquad t \in [0, d).$$

The pair (σ, P) is cyclic if

$$\sigma = \varrho(\sigma, P)$$

and is a ν_ε-approximate cycle if

$$\|\varrho(\sigma, P) - \sigma\| < \nu_\varepsilon.$$

In particular, if $\sigma_0 = (\rho(0), \mathbf{D}^0)$ is the initial state, then the state produced by the process P_t is

$$\sigma(t) = (\rho(t), \mathbf{D}^t)$$

where

$$\rho(t) = \rho(0) \exp\left[-\int_0^t \operatorname{tr}\mathbf{D}(\tau)\, d\tau\right]. \tag{3.7.1}$$

Again we confine the attention to isothermal processes and then we can state the second law of thermodynamics by saying that, for each $\varepsilon > 0$, there exists $\nu_\varepsilon > 0$ such that

$$\int_0^d \hat{\mathbf{T}}(\rho(t), \mathbf{D}^t) \cdot \mathbf{D}(t)\, dt > -\varepsilon$$

for any ν_ε-approximate cycle. Really, we investigate the following law.

Second law for cycles. *For every cycle, the condition*

$$\int_0^d \hat{\mathbf{T}}(\rho(t), \mathbf{D}^t) \cdot \mathbf{D}(t)\, dt \geq 0 \tag{3.7.2}$$

holds; equality holds if and only if the process is reversible, which occurs when $\mathbf{D}^d$ *is the zero history.*

By paralleling the procedure developed for viscoelastic solids, we might show that the validity of the second law for cycles implies the validity of the second law for approximate cycles. That is why we examine (3.7.2) only.

THEOREM 3.7.1. *The constitutive equation* (3.6.2) *complies with the second law for cycles* (3.7.1) *if and only if*

$$\mu_c(\omega) > 0, \qquad \kappa_c(\omega) > 0, \qquad \omega \in \mathbb{R}^{++}. \tag{3.7.3}$$

Proof.[19] Letting H be the integral of $-\Pi(\rho)/\rho$, in view of (3.7.1) we have

$$\int_0^d -\Pi(\rho(t))\, \mathrm{tr}\mathbf{D}(t)\, dt = \int_0^d \frac{dH(\rho(t))}{dt}\, dt.$$

Hence the integral along any cycle of $\Pi(\rho)\mathrm{tr}\mathbf{D}$ vanishes.

Now consider a time-harmonic stretching function $\tilde{\mathbf{D}}(t), t \in \mathbb{R}$, defined as

$$\tilde{\mathbf{D}}(t) = \mathbf{D}_1 \cos\omega t + \mathbf{D}_2 \sin\omega t, \qquad \omega \in \mathbb{R}^{++}, \qquad \mathbf{D}_1, \mathbf{D}_2 \in \mathrm{Sym}.$$

The duration d is taken to be $2\pi/\omega$ times any positive integer. The integral of $\tilde{\mathbf{D}}$ on $[0, d)$ vanishes and hence, by (3.7.1), $\rho(d) = \rho(0)$. Since $\tilde{\mathbf{D}}^d = \tilde{\mathbf{D}}^0$, then $\sigma(t) = (\rho(t), \tilde{\mathbf{D}}^t)$, as t runs over $[0, d)$, constitutes a cycle.

Letting a superposed ring denote the trace-free part, we can write $\mathbf{D}_1 = \mathring{\mathbf{D}}_1 + (1/3)(\mathrm{tr}\mathbf{D}_1)\mathbf{1}$ and the same for $\mathbf{D}_2$. Substitution into (3.7.2) yields

$$2\int_0^d \Big\{ \int_0^\infty \mu(s)[\, \mathring{\mathbf{D}}_1 \cdot \mathring{\mathbf{D}}_1 \cos\omega(t-s) \cos\omega t + \mathring{\mathbf{D}}_2 \cdot \mathring{\mathbf{D}}_2 \sin\omega(t-s) \sin\omega t$$
$$+ \mathring{\mathbf{D}}_1 \cdot \mathring{\mathbf{D}}_2 \,(\cos\omega(t-s) \sin\omega t + \sin\omega(t-s) \cos\omega t)]ds\Big\}dt$$

$$+\int_0^d \Big\{ \int_0^\infty \kappa(s)[(\mathrm{tr}\mathbf{D}_1)^2 \cos\omega(t-s) \cos\omega t + (\mathrm{tr}\mathbf{D}_2)^2 \sin\omega(t-s) \sin\omega t$$
$$+ \mathrm{tr}\mathbf{D}_1 \mathrm{tr}\mathbf{D}_2(\cos\omega(t-s) \sin\omega t + \sin\omega(t-s) \cos\omega t)]ds\Big\}dt > 0$$

whence, by changing the order of integration, we have

$$2\mu_c(\omega)(\mathring{\mathbf{D}}_1 \cdot \mathring{\mathbf{D}}_1 + \mathring{\mathbf{D}}_2 \cdot \mathring{\mathbf{D}}_2) + \kappa_c(\omega)[(\mathrm{tr}\mathbf{D}_1)^2 + (\mathrm{tr}\mathbf{D}_2)^2] > 0, \qquad \omega \in \mathbb{R}^{++},$$

for any nonzero $\mathbf{D}_1, \mathbf{D}_2 \in \mathrm{Sym}$. Hence the results (3.7.3) follow.

To show that (3.7.3) is sufficient for the validity of (3.7.2) along cycles, observe that $(\sigma(0), P)$ on $[0, d)$ constitutes a cycle if and only if the function $\mathbf{D}$ is periodic with vanishing mean value, in the period $[0, d)$. The corresponding history $\mathbf{D}^t$ can be represented through its Fourier series provided only that $\mathbf{D}$ is piecewise continuous. More conveniently we write the Fourier series representation as

$$\mathbf{D}^t(s) = \sum_{h=1}^{\infty} \{\mathring{\mathbf{A}}_h \cos h\omega(t-s) + \mathring{\mathbf{B}}_h \sin h \sin w(t-s)$$
$$+ \tfrac{1}{3}[\mathrm{tr}\mathbf{A}_h \cos h\omega(t-s)$$
$$+ \mathrm{tr}\mathbf{B}_h \sin h\omega(t-s)]\mathbf{1}\}$$

[19]The proof given here is a generalization of the analogous one in [46].

where $\mathbf{A}_h$, $\mathbf{B}_h \in \mathrm{Sym}$, $\omega = 2\pi/d$. We know already that the integral of $p(\rho)\mathrm{tr}\mathbf{D}$ along a cycle vanishes. Then the work w performed along a cycle, namely the left-hand side of (3.7.2), takes the form

$$w = 2\int_0^d \int_0^\infty \mu(s) \sum_{h,k=1}^{\infty} [\mathring{\mathbf{A}}_h \cos h\omega(t-s) + \mathring{\mathbf{B}}_h \sin h\omega(t-s)] \cdot [\mathring{\mathbf{A}}_k \cos k\omega t + \mathring{\mathbf{B}}_k \sin k\omega t] ds\, dt$$
$$+ \int_0^d \int_0^\infty \kappa(s) \sum_{h,k=1}^{\infty} [\mathrm{tr}\mathbf{A}_h \cos h\omega(t-s) + \mathrm{tr}\mathbf{B}_h \sin h\omega(t-s)] [\mathrm{tr}\mathbf{A}_k \cos k\omega t + \mathrm{tr}\mathbf{B}_k \sin k\omega t] ds\, dt.$$

Term by term integration with respect to t gives

$$w = \frac{\pi}{\omega} \sum_{k=1}^{\infty} \{2\mu_c(k\omega)[(\mathring{\mathbf{A}}_k)^2 + (\mathring{\mathbf{B}}_k)^2] + \kappa_c(k\omega)[(\mathrm{tr}\mathbf{A}_k)^2 + (\mathrm{tr}\mathbf{B}_k)^2]\}.$$

The validity of (3.7.3) implies that $w > 0$ for any nonzero history $\mathbf{D}^t$ while $\hat{\mathbf{T}}(\rho, \mathbf{D}^t)$ meets the second law for cycles. □

We may say that the thermodynamic analysis of the viscoelastic fluid (3.6.2) parallels that of the viscoelastic solid (3.1.2). Formally, though, there is a difference in that the power is $\boldsymbol{\mathcal{T}}(\mathbf{E}^t) \cdot \dot{\mathbf{E}}$ for the viscoelastic solid and $\hat{\mathbf{T}}(\rho, \mathbf{D}^t) \cdot \mathbf{D}$ for the viscoelastic fluid. As shown by Theorem 3.7.1, this difference is reflected in the thermodynamic restrictions for the relaxation functions. For, while in viscoelastic solids thermodynamics requires the *negative* definiteness of the half-range Fourier *sine* transform of $\mathbf{G}(s)$, in viscoelastic fluids thermodynamics requires the *positive* definiteness of the half-range Fourier *cosine* transform of $\mu(s)$ and $\kappa(s)$.

To complete this thermodynamic analysis, we have to consider the *incompressible* viscoelastic fluid, namely (3.6.1) or (3.6.2), subject to the constraint

$$\mathrm{tr}\mathbf{D} = 0. \tag{3.7.4}$$

One might think that the incompressible fluid can be regarded as a particular case. It is really so but we have to remark that, while (3.7.4) demands that ρ be constant (a parameter, say), the pressure Π is no longer a function of ρ but is an unknown function of time. The conclusion is given by the following corollary.

COROLLARY 3.7.1. *The viscoelastic incompressible fluid complies with the second law for cycles* (3.7.1) *if and only if*

$$\mu_c(\omega) > 0, \qquad \omega \in \mathbb{R}^{++}. \tag{3.7.5}$$

Proof. The contribution of the pressure Π to the work is

$$-\int_0^d \Pi(t)\,\mathrm{tr}\mathbf{D}(t)\,dt$$

and then it vanishes whatever the dependence of Π on t is. Accordingly, the proof of Theorem 3.7.1 can be applied step by step with the restrictions $\mathrm{tr}\mathbf{A}_h$, $\mathrm{tr}\mathbf{B}_h = 0$ and the result (3.7.5) follows. □

Chapter 4

Existence, Uniqueness, and Stability

4.1. Dynamics of the viscoelastic solid. The thermodynamic restrictions derived in §3.2 for the (linear) viscoelastic solid are essential to the investigation of existence, uniqueness, and stability properties of solutions to the dynamics of the viscoelastic solid. Indeed, this chapter is devoted to the connection between thermodynamic restrictions and the dynamic behaviour of the solid.

There are various ways of looking at the dynamics of the viscoelastic solid, depending on the initial and/or boundary data (cf. Chapter 5) or, rather, the boundary data and the data about the history of the displacement. Here the dynamics of the viscoelastic solid is described through an appropriate Cauchy's problem.

Let $\Omega \subset \mathcal{E}^3$ be a smooth domain occupied by the viscoelastic body in the reference placement and, as usual, let $\mathbf{u} : \Omega \to V$ be the displacement field relative to the reference placement. Cauchy's problem for the viscoelastic solid is expressed as

$$(4.1.1) \qquad \rho\ddot{\mathbf{u}}(\mathbf{x}, t) = \nabla \cdot \mathbf{T}(\mathbf{x}, t) + \rho\mathbf{b}(\mathbf{x}, t) \quad \text{in } \Omega \times \mathbb{R}^{++},$$

$$(4.1.2) \quad \mathbf{T}(\mathbf{x}, t) = \mathbf{G}_0(\mathbf{x})\, \nabla\mathbf{u}(\mathbf{x}, t) + \int_0^\infty \mathbf{G}'(\mathbf{x}, s)\, \nabla\mathbf{u}(\mathbf{x}, t - s)\, ds \quad \text{in } \Omega \times \mathbb{R}^+,$$

$$(4.1.3) \qquad \mathbf{u}(\mathbf{x}, t) = 0 \quad \text{on } \partial\Omega \times \mathbb{R}^+,$$

$$(4.1.4) \qquad \mathbf{u}(\mathbf{x}, t) = \mathbf{u}_0(\mathbf{x}, t) \quad \text{in } \Omega \times \mathbb{R}^-.$$

The relaxation function $\mathbf{G}$ is assumed to be continuous on $\bar{\Omega} \times \mathbb{R}^+$ and to meet the following conditions:
(a) $\mathbf{G} \in C^1(\Omega \times \mathbb{R}^{++}, \text{Lin}(\text{Sym}))$,
(b) $\mathbf{G}'(\mathbf{x}, \cdot) \in L^1(\mathbb{R}^+)$,
(c) $\mathbf{G}_0(\mathbf{x}) > 0$, $\mathbf{x} \in \Omega$.

Remark 4.1.1. We know (cf. §3.1) that $|\mathbf{G}'|^2 k^{-1}$, $k \in L^1(\mathbb{R}^+)$ imply $\mathbf{G}' \in L^1(\mathbb{R}^+)$. Then the fading memory condition (1.2.5) forces (b) to hold. Accordingly, the condition (b) is slightly weaker than the fading memory requirement. In addition, the condition (b) has the advantage of being free from the choice of the influence function k.

On the basis of the conditions (a)–(c), theorems of existence, uniqueness, and continuous dependence on data were proved by Duvaut and Lions [44] and Dafermos [33], [34] for Cauchy's problem (4.1.1)–(4.1.4). The main result is stated as follows.

THEOREM 4.1.1. *If the conditions* (a)–(c) *hold then Cauchy's problem* (4.1.1)–(4.1.4) *with* $\mathbf{b} \in L^2(0, T; L^2(\Omega))$ *and* $\mathbf{u}_0(\mathbf{x}, t)$ *such that*

$$\nabla \cdot \int_t^\infty \mathbf{G}'(\mathbf{x}, s) \nabla \mathbf{u}_0(\mathbf{x}, t - s)\, ds \in L^2(0, T; L^2(\Omega)), \tag{4.1.5}$$

has one and only one weak solution $\mathbf{u} \in H^1(0, T; L^2(\Omega)) \cap L^2(0, T; H_0^1(\Omega))$.

Observe that ρ is a given function of $\mathbf{x} \in \mathcal{R}$. For formal simplicity, in this chapter we let ρ be constant and then we can set $\rho = 1$.

4.2. Ill-posed problems. In the last decade Fichera emphasized, through illuminating papers, that the conditions (a)–(c) are not enough to guarantee the well posedness of Cauchy's problem (4.1.1)–(4.1.4), though Theorem 4.1.1 ensures well posedness in the sense of Hadamard in a finite time interval. Indeed, he exhibited various counterexamples which have addressed the attention to some aspects of the theory neglected up to that time. It is then worth giving an outline of Fichera's counterexamples.

First Fichera [57] considered the one-dimensional relaxation function

$$G(s) = 1 - \lambda + \lambda \exp(-s), \tag{4.2.1}$$

λ being a real-valued parameter. Such relaxation function satisfies the conditions (a)–(c) and then Theorem 4.1.1 applies. Nevertheless, Fichera showed that the spectrum of (4.1.2) is the half space

$$\Re\lambda \geq 1. \tag{4.2.2}$$

This means that for any λ satisfying (4.2.2) we can have nonuniqueness for the solution to the quasi-static problem (see §4.3) in the class $\mathbf{u} \in L^1(\mathbb{R}; H_0^1(\Omega))$.

For a while it was believed[20] that the counterexample was scarcely significant in that $\lambda \geq 1$ corresponds to physically unreasonable forms of the relaxation function because $\lambda > 1$ implies $G(s) < 0$ for sufficiently large values of s while $\lambda = 1$, and then $G_\infty = 0$ corresponds to fluids. Yet, through a second counterexample, Fichera showed that the requirement $G(s) > 0$ is not enough to guarantee existence and uniqueness. Precisely he considered the relaxation function

$$G(s) = \frac{1}{2} - s\,\exp(-s) \tag{4.2.3}$$

for which nonuniqueness is ascertained very easily.[21]

With a view similar to Capriz and Gurtin's, we believed [52] that the counterexample (4.2.3) was scarcely worrisome in that it does not comply with the second law of thermodynamics, notably the inequality (3.2.11).

Setting aside remarks of physical character, the elaboration of these counterexamples by Fichera constituted an important step. It was then clear that, despite the validity of Theorem 4.1.1, even the linear problem (4.1.1)–(4.1.4) might be ill posed. In this regard we mention that the quasi-static problem (cf. §4.3) has been shown to be ill posed [57], [58] and the solution to the dynamic problem (4.1.1)–(4.1.4) to be asymptotically unstable [64].

Soon after the appearance of [52] concerning the compatibility with thermodynamics, Fichera elaborated a third counterexample based on the relaxation function

$$G(s) = G_\infty + (G_0 - G_\infty)\exp(-\lambda s), \tag{4.2.4}$$

where $G_\infty,\ G_0 - G_\infty,\ \lambda \in \mathbb{R}^{++}$, which apparently meets all mechanical and thermodynamic requirements (examined in Chapter 3). Yet the strain function $E(t) = \exp[-(\lambda G_\infty/G_0)t]$ is an eigensolution of

$$G_0 E(t) + \int_0^\infty G'(s)E(t-s)ds = 0$$

thus rendering the solution to (the one-dimensional version of) the quasi-static problem nonunique. Accordingly, the physically well-grounded relaxation function (4.2.4) allows for nonuniqueness if strain functions are considered which are unbounded at the past infinity.

The topic of existence and uniqueness for the solution to various problems involving viscoelastic bodies has been further investigated, especially among

---

[20] See [58] where account is given of a letter to Fichera by Capriz and Gurtin.

[21] Capriz and Gurtin sent the letter to Fichera on May 29, 1980. Fichera answered them immediately by providing the further counterexample (4.2.3) on June 12, 1980.

the Italian researchers on continuum mechanics and related topics (cf., e.g., [72]). Some developments are delineated shortly. Here we observe that, concerning the questions raised by Fichera's counterexamples, we agree on what Fichera wrote to one of us,[22] namely that in essence the difficulties about existence and uniqueness occur because the infinite extent of the memory of the material makes the problem affected by the topology of the space of solutions. These difficulties are such that we can speak of the function space where the problem might be well posed only if the relaxation function is given. Alternatively, given at the outset the function space we can ask for the class of relaxation functions which allow for the well posedness of the problem. In this chapter we let $\mathbf{G}$ be given and look for the function space where the problem is well posed.

The difficulties cannot be removed merely via thermodynamic restrictions. Yet, and this may be seen as the viewpoint underlying this book, we regard the thermodynamic analysis as a preliminary task toward any modelling within continuum mechanics or, more generally, physics. Compatibility of the constitutive properties with thermodynamics does *not* imply per se the well posedness of the possible (dynamic) problems, but addresses the attention to mathematical topics which concern physically sound situations.

4.3. The quasi-static problem. The relaxation functions (4.2.1), (4.2.3), and (4.2.4) have been elaborated as counterexamples to the existence and/or uniqueness of the solution of the so-called quasi-static problem. In such a formulation the inertial term is neglected and then the problem is expressed through the differential equation

$$\nabla \cdot \left[\mathbf{G}_0 \, \nabla \mathbf{u}(t) + \int_0^\infty \mathbf{G}'(s) \, \nabla \mathbf{u}(t-s) \, ds \right] - \mathbf{b}(t) = 0, \tag{4.3.1}$$

in the unknown function $\mathbf{u} : \Omega \times \mathbb{R} \to V$, along with the boundary condition

$$\mathbf{u} = 0 \quad \text{on } \partial\Omega. \tag{4.3.2}$$

The relaxation function $\mathbf{G} : \bar{\Omega} \times \mathbb{R}^+ \to \mathrm{Lin(Sym)}$ is taken to be symmetric,[23] i.e., $\mathbf{G} = \mathbf{G}^T$, and to meet the requirements (a)–(c), and

(d) $\mathbf{G}_\infty > 0$,

(e) $\mathbf{G}'_s(\omega) < 0$, $\omega \in \mathbb{R}^{++}$.

As will be shown in a moment, the physical requirements (d), (e) are not only necessary conditions but, along with (a), (b), and (c), also sufficient for the existence and uniqueness of the solution to a specific problem associated with

---

[22]Letter to A. Morro, on August 1, 1985.

[23]The symmetry condition is not essential to later developments on existence and uniqueness (cf. [91]); it is assumed merely for the sake of formal simplicity.

(4.3.1)–(4.3.2). Hence the viscoelastic solid is here required to satisfy the conditions (a)–(e). According to the literature, stronger monotonicity conditions on the relaxation function are often assumed primarily as necessary conditions to guarantee that (e) holds. Indeed, it seems that no experimental evidence would suggest that stronger monotonicity conditions are physically undesirable. However, (e) is a necessary condition placed by thermodynamics. The next developments show how (e) is also sufficient to guarantee existence and uniqueness of solutions to the quasi-static problem, in the sense of Definition 4.3.1, and the asymptotic stability of the dynamic problem.

To begin with we give the meaning of the term solution.

DEFINITION 4.3.1. A function $\mathbf{u} \in L^2(\mathbb{R}; H_0^1(\Omega))$ is said to be a weak solution to the quasi-static problem (4.3.1)–(4.3.2), with source function $\mathbf{b} \in L^2(\mathbb{R}; L^2(\Omega))$, if $\mathbf{u}$ satisfies (4.3.1)–(4.3.2), in the distributional sense, in $\Omega \times \mathbb{R}$.

Observe that the formal application of the Fourier transform, with respect to t, to (4.3.1) and (4.3.2) provides

(4.3.3)
$$\nabla \cdot [(\mathbf{G}_0(\mathbf{x}) + \mathbf{G}'_F(\mathbf{x},\omega))\nabla \mathbf{u}_F(\mathbf{x},\omega)] = \mathbf{b}_F(\mathbf{x},\omega), \qquad \mathbf{x} \in \Omega, \qquad \omega \in \mathbb{R},$$

(4.3.4)
$$\mathbf{u}_F(\mathbf{x},\omega) = 0 \quad \text{on } \partial\Omega, \qquad \omega \in \mathbb{R}.$$

The subscript F denotes the Fourier transform, namely

$$\mathbf{u}_F(\mathbf{x},\omega) = \int_{-\infty}^{\infty} \mathbf{u}(\mathbf{x},t)\exp(-i\omega t)\,dt.$$

Since $\mathbf{G}(\mathbf{x},\cdot)$ is defined on $\mathbb{R}^+$ only, the Fourier transform $\mathbf{G}'_F$ involves the integral on $\mathbb{R}^+$ instead of $\mathbb{R}$. To save writing, from now on we omit specifying that the conditions (a)–(e) are assumed to hold.

LEMMA 4.3.1. *For any value of ω it is*

(4.3.5)
$$\mathbf{G}_0(\mathbf{x}) + \mathbf{G}'_F(\mathbf{x},\omega) \neq 0, \qquad \mathbf{x} \in \Omega, \qquad \omega \in \mathbb{R}.$$

Proof. Because of (e), when $\omega \neq 0$,

(4.3.6)
$$\Im[\mathbf{G}_0(\mathbf{x}) + \mathbf{G}'_F(\mathbf{x},\omega)] = -\mathbf{G}'_s(\mathbf{x},\omega) \neq 0,$$

and, because of (d), when $\omega = 0$,

(4.3.7)
$$\Re[\mathbf{G}_0(\mathbf{x},0) + \mathbf{G}'_F(\mathbf{x},0)] = \mathbf{G}_0(\mathbf{x}) + \int_0^{\infty} \mathbf{G}'(\mathbf{x},s)\,ds = \mathbf{G}_\infty > 0,$$

which completes the proof. □

In view of (4.3.3) we have to investigate the properties of the operator $L_0(\omega)$, in $H_0^1(\Omega)$, defined as

$$L_0(\omega)\mathbf{u}_F = \nabla \cdot [(\mathbf{G}_0(\mathbf{x}) + \mathbf{G}'_F(\mathbf{x},\omega))\nabla \mathbf{u}_F].$$

In this regard the following result is of decisive importance.

LEMMA 4.3.2. *There exist two constants* $\gamma_1, \gamma_2 \in \mathbb{R}^{++}$, *independent of* ω, *such that, for every* $\mathbf{A} \in \mathrm{Sym}$, *at least one of the inequalities*

$$\mathbf{A}^* \cdot \left[\mathbf{G}_0(\mathbf{x}) + \int_0^\infty \mathbf{G}'(\mathbf{x},s)\cos\omega s\, ds\right]\mathbf{A} \geq \gamma_1\, \mathbf{A}^* \cdot \mathbf{A}, \tag{4.3.8}$$

$$-\mathbf{A}^* \cdot \left[\omega \int \mathbf{G}'(\mathbf{x},s)\sin\omega s\, ds\right]\mathbf{A} \geq \gamma_2\, \mathbf{A}^* \cdot \mathbf{A}, \tag{4.3.9}$$

holds with $\mathbf{A}^*$ *complex conjugate of* $\mathbf{A}$.

Proof. The inequality (e) and the continuity property (a) imply that, at any $\mathbf{x} \in \Omega$, (4.3.9) holds for any ω in the intervals $\omega_0 < |\omega| < \omega_\infty$, with ω_0, ω_∞ strictly positive and finite. Meanwhile, since

$$\lim_{\omega\to\pm\infty} \int_0^\infty \mathbf{G}'(\mathbf{x},s)\cos\omega s\, ds = 0, \tag{4.3.10}$$

$$\lim_{\omega\to 0} \int_0^\infty \mathbf{G}'(\mathbf{x},s)\cos\omega s\, ds = \mathbf{G}_\infty(\mathbf{x}) - \mathbf{G}_0(\mathbf{x}), \tag{4.3.11}$$

upon a suitable choice of ω_0 and ω_∞ the inequality (4.3.8) holds as $|\omega| \in [0,\omega_0]$ and $|\omega| \in [\omega_\infty, \infty)$, too. □

THEOREM 4.3.1. *The solution to the quasi-static problem* (4.3.1)–(4.3.2) *exists and is unique.*

Proof. (Uniqueness). By Lemma 4.3.1 the (vector) equation (4.3.3) turns out to be elliptic and then, for any $\omega \in \mathbb{R}$, the homogeneous problem

$$\nabla \cdot \{[\mathbf{G}_0(\mathbf{x}) + \mathbf{G}'_F(\mathbf{x},\omega)]\nabla \mathbf{u}_F(\mathbf{x},\omega)\} = 0, \qquad \mathbf{x} \in \Omega, \qquad \omega \in \mathbb{R}, \tag{4.3.12}$$

$$\mathbf{u}_F(\mathbf{x},\omega) = 0 \quad \text{on } \partial\Omega, \qquad \omega \in \mathbb{R}, \tag{4.3.13}$$

has the unique solution

$$\mathbf{u}_F(\mathbf{x},\omega) = 0, \qquad \mathbf{x} \in \Omega, \qquad \omega \in \mathbb{R}.$$

This implies that the solution to the problem (4.3.1)–(4.3.2) is unique in that the inverse Fourier transform $\mathbf{u}$ of $\mathbf{u}_F$ necessarily vanishes.

(Existence). To prove the existence of the solution to (4.3.1)–(4.3.2), we need a uniform ellipticity condition for the (real and imaginary) coefficients of (4.3.12). By Lemma 4.3.2, the validity of (4.3.8) or (4.3.9) and the theorems on elliptic equations allow us to say that there exists a unique solution $\mathbf{u}_F(\mathbf{x},\omega) \in H_0^1(\Omega)$, to the problem (4.3.12)–(4.3.13), for any $\omega \in \mathbb{R}$. To complete the proof, we have to show the existence of the solution to the given problem (4.3.1)–(4.3.2).

Consider the function $\mathbf{H}(\mathbf{x},\mathbf{x}',\omega)$, whose values are second-order tensors, such that

$$\nabla' \cdot \{[\mathbf{G}_0(\mathbf{x}) + \mathbf{G}'_F(\mathbf{x},\omega)]\nabla'\mathbf{H}(\mathbf{x},\mathbf{x}',\omega)\} = \delta(\mathbf{x}-\mathbf{x}')\mathbf{1}, \tag{4.3.14}$$

where ∇' stands for the gradient with respect to $\mathbf{x}'$, and

$$\mathbf{H}(\mathbf{x},\mathbf{x}',\omega) = 0 \quad \text{as } \mathbf{x}' \in \partial\Omega. \tag{4.3.15}$$

Again, as a consequence of [120], [56], there exists a unique solution $\mathbf{H}(\mathbf{x},\cdot,\omega) \in H_0^1(\Omega)$ and then we have

$$\mathbf{u}_F(\mathbf{x},\omega) = \int_\Omega \mathbf{H}(\mathbf{x},\mathbf{x}',\omega)\,(\rho\mathbf{b})_F(\mathbf{x}',\omega)\,d\mathbf{x}'. \tag{4.3.16}$$

Moreover, since $\mathbf{G}_0(\mathbf{x}) + \mathbf{G}'_F(\mathbf{x},\omega)$ is continuous in ω, then the function $\mathbf{H}(\mathbf{x},\mathbf{x}',\omega)$ as well is continuous in ω. Then, because $\mathbf{G}'_F(\mathbf{x},\omega) \to 0$ as $\omega \to \pm\infty$, it follows that the Green's function $\mathbf{H}(\mathbf{x},\mathbf{x}',\pm\infty)$ exists and is the solution to the problem

$$\nabla' \cdot [\mathbf{G}_0(\mathbf{x})\nabla'\mathbf{H}(\mathbf{x},\mathbf{x}',\pm\infty)] = \delta(\mathbf{x}-\mathbf{x}'), \tag{4.3.17}$$

$$\mathbf{H}(\mathbf{x},\mathbf{x}',\pm\infty) = 0 \quad \text{as } \mathbf{x}' \in \partial\Omega. \tag{4.3.18}$$

Hence $\mathbf{H}(\mathbf{x},\cdot,\omega) \in H_0^1(\Omega)$ and $\|\mathbf{H}(\mathbf{x},\cdot,\omega)\|_{H_0^1(\Omega)}$ is bounded as $\omega \in \mathbb{R}$. Thus, because $\mathbf{b}_F(\mathbf{x}',\cdot) \in L^2(\mathbb{R})$, by (4.3.16) and the properties of $\mathbf{H}$ we have $\mathbf{u}_F \in L^2(\mathbb{R}; H_0^1(\Omega))$. Then there exists the inverse Fourier transform

$$\mathbf{u}(\mathbf{x},t) = \frac{1}{2\pi}\int_{-\infty}^{\infty} \mathbf{u}_F(\mathbf{x},\omega)\exp(i\omega t)\,d\omega$$

such that $\mathbf{u} \in L^2(\mathbb{R}; H_0^1(\Omega))$ and is the solution to the problem (4.3.1)–(4.3.2). □

In reference to the existence and uniqueness of the solution to the quasi-static problem, it is also worth emphasizing how the hypothesis (e) is essential. The weaker restriction (3.4.1) does not guarantee existence and/or uniqueness of the solution. In fact, if $\mathbf{G}'_s(\omega) \le 0, \omega \in \mathbb{R}^+$, then Lemma 4.3.1 does not apply and $\mathbf{G}_0(\mathbf{x}) + \mathbf{G}'_F(\mathbf{x}, \omega)$ can vanish for some $\omega \in \mathbb{R}^{++}$. Hence the ellipticity condition may cease to hold. This is shown through a family of counterexamples [101], as follows.

Let $\mathbf{C} \in \text{Lin(Sym)}$ be positive definite and possibly dependent on the position $\mathbf{x}$. Then we represent the relaxation function in the form

$$\mathbf{G}(s) = \mathbf{C}\, G(s)$$

and confine the attention to the scalar function $G(s)$ on $\mathbb{R}^+$. Consider the two-parameter family of functions

$$G(s) = \frac{\kappa(\kappa-2)}{16\,\nu^3} - \int_0^s \left(r^2 - \frac{\kappa-1}{\nu} r + \frac{\kappa^2}{8\nu^2}\right) \exp(-\nu r)\, dr, \tag{4.3.19}$$

where $\kappa \in (6,8), \nu > 0$. Of course

$$G(0) = \frac{\kappa(\kappa-2)}{16\,\nu^3}, \qquad G'(s) = -\left(s^2 - \frac{\kappa-1}{\nu} s + \frac{\kappa^2}{8\,\nu^2}\right) \exp(-\nu s).$$

Upon direct evaluation we have

$$G(\infty) = G(0) + \int_0^\infty G'(r)\, dr = -\frac{(\kappa-6)(\kappa-8)}{16\,\nu^3};$$

because $\kappa \in (6,8)$ it is $G(\infty) > 0$. Now we evaluate G'_s to get

$$G'_s(\omega) = -\frac{\omega\kappa^2}{8\nu^2(\nu^2+\omega^2)^3}\left(\omega^2 - \frac{8-\kappa}{\kappa}\nu^2\right)^2$$

whereby $G'_s(\omega) \le 0, \omega \in \mathbb{R}^+$. Indeed, letting

$$\bar{\omega} = \nu\sqrt{\frac{8-\kappa}{\kappa}}$$

we have

$$G'_s(\omega) = 0 \quad \text{as} \quad \omega = 0, \bar{\omega}, \qquad G'_s(\omega) < 0 \quad \text{as} \quad \omega \ne 0, \bar{\omega}.$$

A direct integration yields

$$\int_0^\infty G'(s) \cos \bar{\omega} s\, ds = -\frac{\kappa(\kappa-2)}{16\,\nu^3}$$

and then

$$G(0) + G'_c(\bar{\omega}) = 0.$$

As an aside, we notice that the functions (4.3.19) are not monotone in that $G'(s) > 0$ as s runs between the two (positive) roots of $G'(s)$. However, this does not prevent (4.8.19) from being compatible with thermodynamics.

Now consider the equation of motion, in the quasi-static approximation, with the further assumption that the body force vanishes, namely

$$\nabla \cdot \left\{ \mathbf{C} \left[G(0)\mathbf{E}(\mathbf{x}, t) + \int_0^\infty G'(s)\mathbf{E}(\mathbf{x}, t-s)\, ds \right] \right\} = 0. \tag{4.3.20}$$

Examine the possibility of nonunique solutions to (4.3.20). Evidently, $\mathbf{E}(\mathbf{x}, t) = 0$ is a solution. Let

$$\mathbf{E}(\mathbf{x}, t) = \mathbf{E}_0(\mathbf{x}) \exp(i\bar{\omega}t)$$

with $\mathbf{E}_0$ differentiable but otherwise arbitrary. A nonzero $\mathbf{E}_0$ satisfies (4.3.20) if

$$G(0) \exp(i\bar{\omega}t) + \int_0^\infty G'(s) \exp[i\bar{\omega}(t-s)]\, ds = 0,$$

namely

$$G(0) + \int_0^\infty G'(s) \cos \bar{\omega}s\, ds = 0, \qquad \int_0^\infty G'(s) \sin \bar{\omega}s\, ds = 0.$$

The functions (4.3.19) meet these conditions and hence they allow nonuniqueness of the solution to the quasi-static problem.

4.4. The quasi-static problem with time-harmonic body force. Theorem 4.3.1 deals with body forces $\mathbf{b} \in L^2(\mathbb{R}; L^2(\Omega))$. In many problems, however, the body force $\mathbf{b}$ is oscillatory in time and this dependence sets $\mathbf{b}$ outside $L^2(\mathbb{R}; L^2(\Omega))$. The same occurs, more generally, for forces which do not vanish at infinity. For formal convenience, we begin by investigating the time-harmonic dependence.

Let $\mathbf{b}(\mathbf{x}, t) = \mathbf{b}_0(\mathbf{x}) \exp(i\omega t)$ and look for solutions $\mathbf{u}(\mathbf{x}, t)$ to the quasi-static problem (4.3.1)–(4.3.2) of the form $\mathbf{u}(\mathbf{x}, t) = \mathbf{u}_0(\mathbf{x}) \exp(i\omega t)$. Substitution in (4.3.1) yields

$$\nabla \cdot \left[\mathbf{G}_0(\mathbf{x}) + \int_0^\infty \mathbf{G}'(\mathbf{x}, s) \exp(-i\omega s)\, ds \right] \nabla \mathbf{u}_0(\mathbf{x}) \exp(i\omega t) = \mathbf{b}_0(\mathbf{x}) \exp(i\omega t).$$

Then the quasi-static problem is equivalent to

$$\nabla \cdot \left[\mathbf{G}_0(\mathbf{x}) + \int_0^\infty \mathbf{G}'(\mathbf{x}, s) \exp(-i\omega s)\, ds \right] \nabla \mathbf{u}_0(\mathbf{x}) = \mathbf{b}_0(\mathbf{x}), \tag{4.4.1}$$

(4.4.2) $$\mathbf{u}_0 = 0 \quad \text{on} \quad \partial\Omega,$$

in the unknown function $\mathbf{u}_0 \in H^2(\Omega)$. We know already that $\mathbf{G}_0 + \mathbf{G}'_F \neq 0$ and then that (4.4.1) is elliptic. This proves the following result.

THEOREM 4.4.1. *For any body force* $\mathbf{b}(\mathbf{x},t) = \mathbf{b}_0(\mathbf{x})\exp(i\omega t)$, *with* $\mathbf{b}_0 \in L^2(\Omega)$, *there exists a unique weak solution* $\mathbf{u}(\mathbf{x},t) = \mathbf{u}_0(\mathbf{x})\exp(i\omega t)$, *to the problem* (4.4.1)–(4.4.2), *with* $\mathbf{u}_0 \in H_0^1(\Omega)$.

Now consider the case when, for any position $\mathbf{x}$, the body force $\mathbf{b}(\mathbf{x},t)$ has a finite limit in the past and future infinity. Accordingly, let

$$\lim_{t\to\pm\infty} \mathbf{b}(\mathbf{x},t) = \mathbf{b}_\infty(\mathbf{x}),$$

with $\mathbf{b}_\infty(\mathbf{x}) \in L^2(\Omega)$. Moreover let $\mathbf{u}_\infty(\mathbf{x})$ be the solution to Dirichlet's problem

(4.4.3) $$\nabla \cdot [\mathbf{G}_\infty(\mathbf{x})\nabla \mathbf{u}_\infty(\mathbf{x})] = \mathbf{b}_\infty(\mathbf{x}),$$

(4.4.4) $$\mathbf{u}_\infty(\mathbf{x}) = 0 \quad \text{on } \partial\Omega.$$

Then the functions $\mathbf{u}(\mathbf{x},t)$ are considered which make the difference $\hat{\mathbf{u}}(\mathbf{x},t) = \mathbf{u}(\mathbf{x},t) - \mathbf{u}_\infty(\mathbf{x})$ belong to $L^2(\mathbb{R}; H^2(\Omega))$. Letting $\hat{\mathbf{b}}(\mathbf{x},t) = \mathbf{b}(\mathbf{x},t) - \mathbf{b}_\infty(\mathbf{x})$, look at the problem

(4.4.5) $$\nabla \cdot \left[\mathbf{G}_0(\mathbf{x})\nabla \hat{\mathbf{u}}(\mathbf{x},t) + \int_0^\infty \mathbf{G}'(\mathbf{x},s)\nabla \hat{\mathbf{u}}(\mathbf{x},t-s)\,ds\right] = \hat{\mathbf{b}}(\mathbf{x},t),$$

(4.4.6) $$\hat{\mathbf{u}} = 0 \quad \text{on } \partial\Omega.$$

It is a trivial matter to prove the following theorem.

THEOREM 4.4.2. *Let* $\mathbf{u}_\infty \in H_0^1(\Omega)$ *be the weak solution to the problem* (4.4.3)–(4.4.4), *with* $\mathbf{b}_\infty \in L^2(\Omega)$, *and* $\hat{\mathbf{u}} \in L^2(\mathbb{R}; H_0^1(\Omega))$ *the weak solution to the problem* (4.4.5)–(4.4.6) *with* $\hat{\mathbf{b}} \in L^2(\mathbb{R}; L^2(\Omega))$. *Then* $\mathbf{u}(\mathbf{x},t) = \hat{\mathbf{u}}(\mathbf{x},t) + \mathbf{u}_\infty(\mathbf{x})$ *is the unique weak solution to the problem*

(4.4.7) $$\nabla \cdot \left[\mathbf{G}_0(\mathbf{x})\nabla \mathbf{u}(\mathbf{x},t) + \int_0^\infty \mathbf{G}'(\mathbf{x},s)\nabla \mathbf{u}(\mathbf{x},t-s)\,ds\right] = \mathbf{b}(\mathbf{x},t),$$

(4.4.8) $$\mathbf{u} = 0 \quad \textit{on} \quad \partial\Omega,$$

where $\mathbf{b}(\mathbf{x},t) = \hat{\mathbf{b}}(\mathbf{x},t) + \mathbf{b}_\infty(\mathbf{x})$.

Proof. The positive definiteness of $\mathbf{G}_\infty(\mathbf{x})$ makes the equation (4.4.3) elliptic and gives Dirichlet's problem (4.4.3)–(4.4.4) a unique solution $\mathbf{u}_\infty \in H_0^1(\Omega)$. Now, the problem (4.4.5)–(4.4.6), for $\hat{\mathbf{u}}$, is formally the same as the problem (4.4.1)–(4.4.2) for which we have proved, through Theorem 4.3.1, existence and uniqueness of the solution in $L^2(\mathbb{R}; H_0^1(\Omega))$. Then there exists a unique solution $\mathbf{u} = \mathbf{u}_\infty + \hat{\mathbf{u}}$, $\hat{\mathbf{u}} \in L^2(\mathbb{R}; H_0^1(\Omega))$, to the problem (4.4.7)–(4.4.8). □

4.5. The dynamic problem. Really the dynamics of a viscoelastic body is governed by (4.1.1) and (4.1.2) (in the case $\rho = 1$); relative to the quasi-static problem examined so far we have to account also for the inertial term. Accordingly, the remaining part of this chapter is devoted to the investigation of the dynamic problem

(4.5.1)
$$\ddot{\mathbf{u}}(\mathbf{x},t) = \nabla\cdot\left[\mathbf{G}_0(\mathbf{x})\nabla\mathbf{u}(\mathbf{x},t) + \int_0^\infty \mathbf{G}'(\mathbf{x},s)\nabla\mathbf{u}(t-s)\,ds\right] + \mathbf{b}(\mathbf{x},t) \quad \text{on } \Omega\times\mathbb{R},$$

(4.5.2)
$$\mathbf{u}(\mathbf{x},t) = 0 \quad \text{on } \partial\Omega\times\mathbb{R}.$$

To such equations one should adjoin the initial conditions as in (4.1.4). Alternatively, one can make suitable hypotheses on $\mathbf{b}$ and look for the properties of the solution $\mathbf{u}(\mathbf{x},t)$ with $t \in \mathbb{R}$. It is just this type of problem that is investigated in this section.

Incidentally, besides being of interest on its own, the dynamic problem may be connected to the questions raised by Fichera's counterexamples in that, as conjectured by Capriz [9], nonexistence and nonuniqueness might originate from the quasi-static approximation. By accounting for the inertial term as well (dynamic problem), some, if not all, drawbacks might disappear. With this view, a research has recently been undertaken by Virga and Capriz [124] which is outlined below.

4.5.1. The displacement decays exponentially in the past. Consider the one-dimensional version of (4.5.1) with $\Omega = [0,l], l > 0$. Let $G \in C^2(\mathbb{R}^+,\mathbb{R}^+)$. Moreover let $u(x,t) \in C^2(\Sigma)$, with

$$\Sigma = \{(x,t); x \in [0,l], t \in \mathbb{R}^-\},$$

and

$$u(0,t) = u(l,t) = 0, \qquad t \in \mathbb{R}^-.$$

The counterpart of the initial data is expected to be a suitable restriction on the behaviour of the solution at the past infinity. Owing to linearity, we can say that the solution to (4.5.1) is unique if and only if there exists only the zero solution to (4.5.1) when $\mathbf{b}$ vanishes.

Given $\varepsilon \in \mathbb{R}^{++}$ consider the Banach spaces

$$U_\varepsilon = \{u \in C(\Sigma) : \sup \exp(-\varepsilon t)|u(x,t)| < \infty\},$$

$$V_\varepsilon = \{v \in C(\mathbb{R}^-) : \sup \exp(-\varepsilon t)|v(t)| < \infty\},$$

$$V'_\varepsilon = \left\{ v \in C(\mathbb{R}^+) : \int_0^\infty |v(t)| \exp(-\varepsilon t)\, dt < \infty \right\}$$

with the norms $\|u\|_\varepsilon$, $\|v\|_\varepsilon$, $\|v\|'_\varepsilon$ as the supremums and the integral appearing in the definition of U_ε, V_ε, V'_ε, respectively. For later purposes it is worth proving the following lemma.

LEMMA 4.5.1. *If f is absolutely integrable on $\mathbb{R}^-$, then the differential equation*

$$\alpha^2 \ddot{v}(t) + \beta^2 v(t) = f(t), \qquad t \in \mathbb{R}^-, \qquad \alpha, \beta \neq 0, \tag{4.5.3}$$

has only one solution of class C^2 such that $v, \dot{v} \in V_\varepsilon$, that is,

$$v(t) = \frac{1}{\alpha\beta} \int_{-\infty}^t \sin \frac{\beta}{\alpha}(t-\tau)\, f(\tau)\, d\tau. \tag{4.5.4}$$

Proof. For any $s \in (-\infty, t)$ we can write the solution to (4.5.3) as

$$v(t) = v(s) \cos \frac{\beta}{\alpha}(t-s) + \frac{\alpha}{\beta} \sin \frac{\beta}{\alpha}(t-s) + \frac{1}{\alpha\beta} \int_s^t \sin \frac{\beta}{\alpha}(t-\tau)\, f(\tau)\, d\tau.$$

The fact that $v, \dot{v} \in V_\varepsilon$ implies that $v(s) \to 0$ as $s \to -\infty$. Then the limit $s \to -\infty$ provides the solution in the form (4.5.4). □

We are now in a position to prove the desired uniqueness result.

THEOREM 4.5.1. *Let $G(x, \cdot)$ be such that $G', G'' \in V'_\varepsilon$ and*

(4.5.5)

$$G_0 > \int_0^\infty |G'(s)| \exp(-\varepsilon s)\, ds + \frac{1}{\varepsilon}\left(|G'_0| + \int_0^\infty |G''(s)| \exp(-\varepsilon s)\, ds \right).$$

Then the solution to the problem (4.5.1)–(4.5.2) is unique.

Proof. Let $u \in C^2(\Sigma)$. For any $t \in \mathbb{R}^-$ extend $u(\cdot, t)$ to $[-l, l]$ by letting

$$\bar{u}(x,t) = \begin{cases} u(x,t), & x \in (0,l], \\ -u(x,t), & x \in [-l, 0); \end{cases}$$

$\bar{u}$ is of class C^2 in $\bar{\Sigma} = \{(x,t) : |x| \in (0,l], t \in \mathbb{R}^-\}$. We require $\bar{u}$ to be solution to the homogeneous problem

$$-\bar{u}_{tt}(x,t) + G_0 \bar{u}_{xx}(x,t) + \int_{-\infty}^t G'(t-\tau) \bar{u}_{xx}(x,\tau)\, d\tau = 0, \tag{4.5.6}$$

$$\bar{u}(-l,t) = \bar{u}(0,t) = \bar{u}(l,t) = 0, \qquad t \in \mathbb{R}^-,$$

the subscripts t, x denoting partial derivatives. Let

$$v_n(t) = \frac{1}{l}\int_0^l u(x,t)\,\sin\frac{n\pi x}{l}\,dx, \qquad t \in \mathbb{R}^-,$$

with $n = 1, 2, \cdots$; of course $v_n \in C^2(\mathbb{R}^{--})$ for any n. Moreover $u, u_t \in U_\varepsilon$ implies $v, \dot{v} \in V_\varepsilon$. Multiplication of (4.5.6) by $\sin(n\pi x/l)$ and integration gives

$$\left(\frac{l}{n\pi}\right)^2 \ddot{v}_n(t) + G_0 v_n(t) = -\int_{-\infty}^t G'(t-\tau)v_n(\tau)\,d\tau, \qquad t \in \mathbb{R}^-.$$

Then, letting

$$\alpha = \frac{l}{n\pi}, \qquad \beta = \sqrt{G_0}, \qquad f(t) = -\int_{-\infty}^t G'(t-\tau)\,v_n(\tau)\,d\tau$$

and applying Lemma 4.5.1, we have

$$v_n(t) = -\frac{\lambda_n}{G_0}\int_{-\infty}^t d\sigma\,\sin\lambda_n(t-\sigma)\int_{-\infty}^\sigma G'(\sigma-\tau)\,v_n(\tau)\,d\tau,$$

where

$$\lambda_n = \frac{n\pi}{l}\sqrt{\frac{G_0}{\rho}}.$$

Since $G'' \in V'_\varepsilon$ an integration by parts yields

$$v_n = \mathcal{F}_n(v_n)$$

where $\mathcal{F}_n$ is the linear operator on V_ϵ defined by

$$\begin{aligned}\mathcal{F}_n(v) = \frac{1}{G_0}\Bigg[&-\int_{-\infty}^t G'(t-\tau)\,v(\tau)\,d\tau \\ &+ G'(0)\int_{-\infty}^t \cos\lambda_n(t-\tau)\,v(\tau)\,d\tau \\ &+ \int_{-\infty}^t d\sigma\cos\lambda_n(t-\sigma)\int_{-\infty}^\sigma G''(\sigma-\tau)\,v(\tau)\,d\tau\Bigg].\end{aligned}$$

Hence it follows that, for any $t \in \mathbb{R}^{--}$,

$$\begin{aligned}\exp(-\varepsilon t)|\mathcal{F}_n(v)| \le \frac{1}{G_0}\Bigg[&\int_0^\infty |G'(s)|\exp(-\varepsilon s)\,ds \\ &+ \left(|G'(0)| + \int_0^\infty |G''(s)|\exp(-\varepsilon s)\,ds\right)\int_0^\infty \exp(-\varepsilon s)|\cos\lambda_n(t-s)|\,ds\Bigg]\|v\|_\varepsilon.\end{aligned}$$

By virtue of (4.5.5) $\mathcal{F}_n$ turns out to be a contraction mapping in V_ε and then v_n vanishes identically, for each n, in $\mathbb{R}^{--}$. $\square$

As an example, look at the relaxation function (4.2.4.), namely

$$G(s) = G_\infty + (G_0 - G_\infty)\exp(-\lambda s), \qquad G_\infty,\ G_0 - G_\infty,\ \lambda > 0,$$

which satisfies the constitutive requirements (a)–(e). A trivial calculation shows that the condition (4.5.5) holds for

$$\frac{\lambda}{\varepsilon} \in \Big(0, \frac{G_0}{2(G_0 - G_\infty)}\Big).$$

This example shows that, according to the condition (4.5.5), counterexamples to uniqueness can be elaborated, once G is given, by choosing ε sufficiently small; of course, we let

$$|G_0'| + \int_0^\infty |G''(s)|\exp(-\varepsilon s)\,ds > 0.$$

As a by-product, elaborating a counterexample in the dynamic case may shed light on the connection with the quasi-static approximation.

Consider again the relaxation function (4.2.4) and look for solutions of the form

$$u(x,t) = u_0 \sin\frac{\pi x}{l}\,\exp[(\mu + i\nu)t]$$

where $\mu,\ \nu \in \mathbb{R}$. Substitution into (4.5.6) gives

$$-\frac{l^2}{\pi^2}(\mu^2 - \nu^2) + G_0 - (G_0 - G_\infty)\frac{\lambda(\lambda+\mu)}{(\lambda+\mu)^2 + \nu^2}$$

and

$$-2\frac{l^2}{\pi^2}\mu\nu + (G_0 - G_\infty)\frac{\lambda\nu}{(\lambda+\mu)^2+\nu^2} = 0.$$

In the quasi-static case (formally, $\rho = 0$) we have

$$\nu = 0, \qquad \mu = -\frac{G_\infty}{G_0}\lambda,\ -\lambda,$$

one being just the third counterexample provided by Fichera. Since $\lambda > 0$, this means that eigensolutions are unbounded at past infinity. In the dynamic case, instead, we have $\mu > 0, \nu \neq 0$.

A suggestive way of elaborating a counterexample was adopted in [10] by letting ε be a small parameter and considering

$$\nu = \nu_0 + \varepsilon\nu_1 + o(\varepsilon), \qquad \mu = \varepsilon\mu_1 + o(\varepsilon).$$

Furthermore, the density was taken as $\rho = \rho_0\varepsilon$, thus regarding the density as the perturbative parameter. Upon substitution, at the zero order and first order in ε we have

$$G_0 + \int_0^\infty G'(s)\cos\nu_0 s\,ds = 0,$$

$$\int_0^\infty G'(s)\sin\nu_0 s\,ds = 0,$$

and

$$\rho_0\frac{l^2}{\pi^2}\nu_o^2 + (\mu_1 + i\nu_1)\int_0^\infty sG'(s)\exp(-i\nu_0 s)\,ds = 0.$$

Then μ_1 is allowed to be negative provided that

$$\int_0^\infty s\,G'(s)\cos\nu_0 s\,ds < 0.$$

This is viewed as the condition ensuring the existence of counterexamples.[24] Capriz and Virga considered some examples of relaxation functions compatible with this condition, such as

$$G(s) = (G_\infty + 2G_0)\exp(-2\lambda s) - (2G_\infty + G_0)\exp(-\lambda s),$$
$$\lambda > 0, \qquad 4G_\infty > G_0 > 0,$$

with $\nu_0 = \pm\lambda\sqrt{2G_\infty/G_0}$. However, they were unable to determine relaxation functions compatible with the thermodynamic requirement (e), namely $G'_s(\omega) < 0,\ \omega \in \mathbb{R}^{++}$.

4.5.2. The Fourier transform of the displacement is in H_0^1. Back to the three-dimensional case, now we follow [91] and prove an existence and uniqueness result. By analogy with the quasi-static problem, consider the operator

$$L(\omega)\mathbf{u}_F = \omega^2\mathbf{u}_F + \nabla\cdot[(\mathbf{G}_0(\mathbf{x}) + \mathbf{G}'_F(\mathbf{x},\omega))\nabla\mathbf{u}_F]$$

in H_0^1. Upon the application of the Fourier transform, the problem (4.5.1)–(4.5.2) leads to

$$L(\omega)\mathbf{u}_F(\mathbf{x},\omega) = \mathbf{b}_F(\mathbf{x},\omega), \tag{4.5.7}$$

$$\mathbf{u}_F(\mathbf{x},\omega) = 0 \quad \text{on } \partial\Omega. \tag{4.5.8}$$

As with $L_0(\omega)$, $L(\omega)$ is uniformly coercive in $H_0^1(\Omega)$ in the following sense.

---

[24]Unfortunately the procedure does not deliver the possible explicit solution for μ and ν.

LEMMA 4.5.2. *There exist two constants $\nu_1, \nu_2 \in \mathbb{R}^{++}$, independent of ω, such that at least one of the inequalities*

$$\int_\Omega \{\nabla \mathbf{u}_F^* \cdot [\mathbf{G}_0(\mathbf{x}) + \mathbf{G}_c'(\mathbf{x},\omega)]\nabla \mathbf{u}_F + \omega^2 \mathbf{u}_F^* \cdot \mathbf{u}_F\}dx \geq \nu_1 \int_\Omega (|\nabla \mathbf{u}_F|^2 + |\mathbf{u}_F|^2)dx, \tag{4.5.9}$$

$$-\omega \int_\Omega \nabla \mathbf{u}_F^* \cdot \mathbf{G}_s'(\mathbf{x},\omega)\nabla \mathbf{u}_F \, dx \geq \nu_2 \int_\Omega (|\nabla \mathbf{u}_F|^2 + |\mathbf{u}_F|^2)dx, \tag{4.5.10}$$

holds.

Proof. By virtue of Lemma 4.3.2, the application of Poincaré's inequality and the observation that $\mathbf{u}_F \in H_0^1(\Omega)$, implies the validity of (4.5.10) for $|\omega| \in (\omega_0, \omega_\infty)$ with ω_0, ω_∞ strictly positive and finite. Now we prove that (4.5.9) holds as ω is close or equal to zero. When $\omega = 0$ the left-hand side of (4.5.9) reduces to

$$\int_\Omega \nabla \mathbf{u}_F^* \cdot \mathbf{G}_\infty(\mathbf{x}) \, \nabla \mathbf{u}_F \, dx.$$

Then the positive definiteness of $\mathbf{G}_\infty$ allows us to write

$$\int_\Omega \nabla \mathbf{u}_F^* \cdot \mathbf{G}_\infty(\mathbf{x}) \, \nabla \mathbf{u}_F \, dx \geq \nu_0 \int_\Omega |\nabla \mathbf{u}_F|^2 dx,$$

for a suitable ν_0. Again we have recourse to Poincaré's inequality and, owing to the continuity of $\mathbf{G}_c'(\mathbf{x},\cdot)$, we obtain (4.5.9) for $|\omega| \leq \omega_0$. Finally, by means of (4.3.10) and the positive definiteness of $\mathbf{G}_0$ we obtain (4.5.9) for $|\omega| \geq \omega_\infty$. □

To prove the existence of a weak solution to the problem (4.5.7)–(4.5.8), it is convenient to examine preliminarily the problem

$$\omega^2 \mathbf{H}(\mathbf{x},\mathbf{x}',\omega) - \nabla' \cdot \{[\mathbf{G}_0(\mathbf{x}') + \mathbf{G}_F'(\mathbf{x}',\omega)]\nabla' \mathbf{H}(\mathbf{x},\mathbf{x}',\omega)\} = \delta(\mathbf{x}-\mathbf{x}')\mathbf{1}, \tag{4.5.11}$$

$$\mathbf{H}(\mathbf{x},\mathbf{x}',\omega) = 0 \quad \text{as } \mathbf{x}' \in \partial\Omega, \tag{4.5.12}$$

in the unknown Green's tensor function $\mathbf{H}(\mathbf{x},\mathbf{x}',\omega)$; in indicial notation $[\nabla'(\mathbf{G}_0 \nabla' \mathbf{H})]_{ij} = \partial[(G_0)_{ihlm} \partial H_{mj}/\partial x_l']/\partial x_h'$.

LEMMA 4.5.3. *There exists a unique solution $\mathbf{H}(\mathbf{x},\mathbf{x}',\omega)$ to the problem (4.5.11)–(4.5.12) such that $\mathbf{H}(\mathbf{x},\cdot,\omega) \in H_0^1(\Omega)$, $\mathbf{x} \in \Omega$; moreover $\mathbf{H}(\mathbf{x},\mathbf{x}',\cdot)$ is continuous and bounded with $\mathbf{H}(\mathbf{x},\mathbf{x}',\omega) = O(\omega^{-2+\varepsilon})$ as $\omega \to \infty$.*

Proof. We know that $L(\omega)$ is a coercive operator in $H_0^1(\Omega)$. Then by Lax-Milgram's theorem $L(\omega)$ is an isomorphism from $H_0^1(\Omega)$ onto $H^{-1}(\Omega)$ (cf.

[120], Lemma 23.1). Since $\delta \in H^{-1}(\Omega)$ then the tensor-valued function $\mathbf{H}$ exists, is unique, and belongs to $H_0^1(\Omega)$.

By the continuous dependence of $L(\omega)$ on ω, application of [120, Lemma 44.1] provides the continuous dependence of $\mathbf{H}$ on ω.

Now examine the behaviour of $\mathbf{H}$ as $\omega \to \infty$. Let $\phi \in C_0^\infty(\Omega)$ be real-valued. By (4.5.11) we have

$$\int_\Omega \mathbf{H}(\mathbf{x},\mathbf{x}',\omega)\phi(\mathbf{x}')\,dx'$$
$$= \frac{1}{\omega^2}\left\{\int_\Omega \nabla'\cdot[(\mathbf{G}_0(\mathbf{x}')+\mathbf{G}_F'(\mathbf{x}',\omega))\nabla'\mathbf{H}(\mathbf{x},\mathbf{x};,\omega)]\phi(\mathbf{x}')\,dx' + \phi(\mathbf{x})\right\}.$$

Application of the divergence theorem yields

$$\int_\Omega \mathbf{H}(\mathbf{x},\mathbf{x}',\omega)\left\{\phi(\mathbf{x}')-\frac{1}{\omega^2}\nabla'\cdot[(\mathbf{G}_0(\mathbf{x}')+\mathbf{G}_F'(\mathbf{x}',\omega))\nabla'\phi(\mathbf{x}')]\right\}dx' = \frac{1}{\omega^2}\phi(\mathbf{x}).$$

Hence we have

$$\lim_{\omega\to\infty}\omega^{2-\varepsilon}\int_\Omega \mathbf{H}(\mathbf{x},\mathbf{x}',\omega)\phi(\mathbf{x}')\,dx' = \omega^{-\varepsilon}\int_\Omega \nabla'\cdot[\mathbf{G}_0(\mathbf{x}')\nabla'\phi(\mathbf{x}')]\,dx' + \phi(\mathbf{x}).$$

The observation that the function ϕ is arbitrary completes the proof. □

We are now in a position to prove the main result.

THEOREM 4.5.2. *For every body force* $\mathbf{b} \in L^2(\mathbb{R}; L^2(\Omega))$ *there exists a unique solution*

$$\mathbf{u} \in L^2(\mathbb{R}; H_0^1(\Omega)) \cap H^1(\mathbb{R}; L^2(\Omega))$$

to the problem (4.5.1)–(4.5.2).

Proof. We know that, upon Fourier transformation, the problem (4.5.1)–(4.5.2) leads to (4.5.7)–(4.5.8) in the unknown function $\mathbf{u}_F$. By Lemma 4.5.3, $\mathbf{H}(\mathbf{x},\mathbf{x}',\omega)$ is the unique solution to the problem (4.5.11)–(4.5.12) with $\mathbf{H}(\mathbf{x},\cdot,\omega) \in H_0^1(\Omega)$ while $\mathbf{H}(\mathbf{x},\mathbf{x}',\cdot)$ is bounded and continuous and $\mathbf{H}(\mathbf{x},\mathbf{x}',\omega) = O(\omega^{-2+\varepsilon})$ as $\omega \to \infty$. Then we can write the solution $\mathbf{u}_F$ to (4.5.7)–(4.5.8) as

$$\mathbf{u}_F(\mathbf{x},\omega) = \int_\Omega \mathbf{H}(\mathbf{x},\mathbf{x}',\omega)\,\mathbf{b}_F(\mathbf{x}',\omega)\,dx'. \tag{4.5.13}$$

Since $\mathbf{b} \in L^2(\mathbb{R}; L^2(\Omega))$ then we have $\mathbf{b}_F \in L^2(\mathbb{R}; L^2(\Omega))$. Accordingly there exists the inverse Fourier transform of the function (4.5.13), that is,

$$\mathbf{u}(\mathbf{x},t) = \frac{1}{2\pi}\int_{-\infty}^{\infty}\mathbf{u}_F(\mathbf{x},\omega)\exp(-i\omega t)\,d\omega$$
$$= \frac{1}{2\pi}\int_\Omega\int_{-\infty}^{\infty}\mathbf{H}(\mathbf{x},\mathbf{x}',\omega)\,\mathbf{b}_F(\mathbf{x}',\omega)\exp(-i\omega t)\,d\omega\,dx',$$

which proves the existence of the solution to the problem (4.5.1)–(4.5.2).

As regards uniqueness, let $\mathbf{u}_1, \mathbf{u}_2$ be two solutions to (4.5.1)–(4.5.2) corresponding to the same body force $\mathbf{b} \in L^2(\mathbb{R}; L^2(\Omega))$. The Fourier transform $\mathbf{u}_F$ of the difference $\mathbf{u} = \mathbf{u}_1 - \mathbf{u}_2$ is a solution to the homogeneous problem

$$L(\omega)\,\mathbf{u}_F(\mathbf{x},\omega) = 0, \qquad \mathbf{u}_F(\mathbf{x},\omega) = 0 \quad \text{on } \partial\Omega. \tag{4.5.14}$$

The analogue of (4.5.13) shows that $\mathbf{u}_F(\mathbf{x},\omega) = 0$ is the unique solution to (4.5.14). Upon inverse Fourier transforming, we conclude that $\mathbf{u}(\mathbf{x},t) = 0$ is the unique solution to the homogeneous problem for $\mathbf{u}$. □

4.6. Cauchy's problem: existence, uniqueness, and stability. In this section we go back to Cauchy's problem with Dirichlet's conditions, as expressed by (4.1.1)–(4.1.4), and investigate the properties of the solution $\mathbf{u}$ in the space-time domain $\Omega \times \mathbb{R}^+$.

The relaxation function $\mathbf{G}$ is still required to satisfy the conditions (a)–(e) of §§4.1 and 4.3. This allows us to prove the following, preliminary result [63].

LEMMA 4.6.1. *If* $\mathbf{G}$ *satisfies* (3.1.1) *and* (3.2.11) *then, for every pair of real parameters* α, β, *with* $\alpha \geq 0$ *and* $\beta \neq 0$, *there exist two positive-valued functions* $\gamma_1(\alpha)$ *and* $\gamma_2(\alpha,\beta)$ *such that, for each* $\mathbf{x} \in \Omega$ *and* $\mathbf{A} \in \mathrm{Sym}$,

$$\mathbf{A} \cdot \left[\mathbf{G}_0(\mathbf{x}) + \int_0^\infty \exp(-\alpha s)\,\mathbf{G}'(\mathbf{x},s)\,ds\right]\mathbf{A} \geq \gamma_1(\alpha)\mathbf{A}\cdot\mathbf{A}, \tag{4.6.1}$$

$$\mathbf{A} \cdot \Big\{-\beta \int_0^\infty \exp(-\alpha s)\sin\beta s\,\mathbf{G}'(\mathbf{x},s)\,ds\Big\}\mathbf{A} \geq \gamma_2(\alpha,\beta)\mathbf{A}\cdot\mathbf{A}. \tag{4.6.2}$$

Proof. By Theorem 3.4.2 and the continuity of $\mathbf{G}'$ (property (a) of §4.1) we obtain at once (4.6.1). By the same token, Theorem 3.4.1 yields (4.6.2). □

Now we improve previous results in that, besides proving existence and uniqueness, we show that stability properties hold. This is accomplished by letting

$$\mathbf{b} \in L^2(\mathbb{R}^+; H^{-1}(\Omega)), \tag{4.6.3}$$

$$\mathbf{u}_0(\cdot,\tau) \in H_0^1(\Omega), \qquad \dot{\mathbf{u}}_0(\cdot,\tau) \in L^2(\Omega), \qquad \tau \in \mathbb{R}^-, \tag{4.6.4}$$

and, moreover, the initial history $\mathbf{u}_0$ be such that the function

$$\mathbf{U}(\mathbf{x},t) = \nabla \cdot \int_t^\infty \mathbf{G}'(\mathbf{x},s)\nabla\mathbf{u}_0(\mathbf{x},t-s)\,ds \in L^2(\mathbb{R}^+, H^{-1}(\Omega)), \tag{4.6.5}$$

namely

$$\int_0^\infty \left| \int_\Omega \left[\int_t^\infty \mathbf{G}'(\mathbf{x}, s) \nabla \mathbf{u}_0(\mathbf{x}, t-s)\, ds \right] \cdot \nabla \mathbf{v}(\mathbf{x}) \right|^2 dt < \infty, \qquad \forall \mathbf{v} \in H_0^1(\Omega).$$

To study the stability properties of the solution $\mathbf{u}$, it is convenient to investigate the behaviour of the Laplace transform $\mathbf{u}_L(p) = \int_0^\infty \exp(-pt)\mathbf{u}(t)\, dt$ of $\mathbf{u}$, p being a complex-valued parameter. Application of the Laplace transform operator to (4.1.1)–(4.1.4) leads to

$$\begin{aligned} p^2\mathbf{u}_L(\mathbf{x},p) &- \nabla \cdot \{[\mathbf{G}_0(\mathbf{x}) + \mathbf{G}'_L(\mathbf{x},p)]\nabla \mathbf{u}_L(\mathbf{x},p)\} \\ &= \mathbf{b}_L(\mathbf{x},p) + p\,\mathbf{u}(\mathbf{x},0) + \dot{\mathbf{u}}(\mathbf{x},0) + \mathbf{U}_L(\mathbf{x},p) =: \mathbf{B}_L(\mathbf{x},p), \end{aligned} \tag{4.6.6}$$

$$\mathbf{u}_L(\mathbf{x},p) = 0 \quad \text{on } \partial\Omega \tag{4.6.7}$$

where $\mathbf{U}_L(\mathbf{x},p)$ is the Laplace transform of $\mathbf{U}(\mathbf{x},t)$. We have the following result.

LEMMA 4.6.2. *If* $\mathbf{b}$ *and* $\mathbf{u}_0$ *satisfy* (4.6.3)–(4.6.5) *then there exists a unique solution to the problem* (4.6.6)–(4.6.7) *in* $H_0^1(\Omega)$ *for every* $p \in \mathbb{C}^+$.

Proof. Since $\mathbf{b} \in L^2(\mathbb{R}^+; H^{-1}(\Omega))$ and $\mathbf{u}_0$ satisfies (4.6.4), (4.6.5), the Laplace transforms $\mathbf{b}_L(\mathbf{x},p)$ and $\mathbf{U}_L(\mathbf{x},p)$ are well defined for every $p \in \mathbb{C}^{++}$. If, though, $p = i\beta$, $\beta \in \mathbb{R}$, then $\mathbf{b}_L(\mathbf{x}, i\beta)$ and $\mathbf{U}_L(\mathbf{x}, i\beta)$ can be considered as L^2-Fourier transforms of $\mathbf{b}$ and $\mathbf{u}_0$. Accordingly we conclude that the Laplace transforms $\mathbf{b}_L(\mathbf{x},p)$ and $\mathbf{U}_L(\mathbf{x},p)$ are well defined for every $p \in \mathbb{C}^+$.

Following known theorems on elliptic equations [56], to prove the desired result we need only show that the operator

$$L(p)\mathbf{u}_L = -p^2\mathbf{u}_L - \nabla \cdot \{[\mathbf{G}_0(\mathbf{x}) + \mathbf{G}_L(\mathbf{x},p)]\nabla \mathbf{u}_L\}$$

is coercive in that the bilinear form

$$a(\mathbf{u},\mathbf{v};p) = \int_\Omega \{p^2\mathbf{u}(\mathbf{x}) \cdot \mathbf{v}^*(\mathbf{x}) + \nabla \mathbf{v}^*(\mathbf{x}) \cdot [\mathbf{G}_0(\mathbf{x}) + \mathbf{G}'_L(\mathbf{x},p)]\nabla \mathbf{u}(\mathbf{x})\} dx$$

satisfies the inequality

$$|a(\mathbf{u},\mathbf{u};p)| > k(p)\|\mathbf{u}\|^2_{H_0^1(\Omega)}$$

where $k(p)$ is a strictly positive constant, for every $\mathbf{u} \in H_0^1(\Omega)$ and $p \in \mathbb{C}^+$. Let, as usual, $p = \alpha + i\beta$ and look first at the case $\beta = 0$. We have

$$\mathbf{G}_0(\mathbf{x}) + \mathbf{G}'_L(\mathbf{x},\alpha) = \mathbf{G}_0(\mathbf{x}) + \int_0^\infty \exp(-\alpha s)\, \mathbf{G}'(\mathbf{x},s)\, ds,$$

and hence the bilinear form becomes

$$a(\mathbf{u}, \mathbf{u}; \alpha) = \int_\Omega \{\alpha^2 \mathbf{u} \cdot \mathbf{u}^* + \nabla \mathbf{u} \cdot [\mathbf{G}_0(\mathbf{x}) + \mathbf{G}'_L(\mathbf{x}, \alpha)] \nabla \mathbf{u}^* \} dx.$$

Then by the positive definiteness expressed by (4.6.1) and Poincaré's inequality, we have

$$|a(\mathbf{u}, \mathbf{u}, \alpha)| > \nu(\alpha) \|\mathbf{u}\|^2_{H^1_0(\Omega)},$$

where $\nu(\alpha) > 0$ as $\alpha \in \mathbb{R}^+$.

Now let $\beta \neq 0$. It follows at once that

$$\beta \Im a(\mathbf{u}, \mathbf{u}, p) = \int_\Omega (2\,\alpha\beta^2 \mathbf{u} \cdot \mathbf{u}^* + \beta \nabla \mathbf{u}^* \cdot \Im \mathbf{G}_L \nabla \mathbf{u}) dx.$$

By (4.6.2) we can write

$$|\Im a(\mathbf{u}, \mathbf{u}; p)| \geq \gamma_2(p) \Big(\int_\Omega |\nabla \mathbf{u}(\mathbf{x})|^2 \, dx + \int_\Omega |\mathbf{u}(\mathbf{x})|^2 \, dx \Big)$$

where $\gamma_2(p) > 0$ as $\beta \neq 0$. Hence there exists a positive-valued function $\gamma(p)$ such that

$$|a(\mathbf{u}, \mathbf{u}; p)| \geq \gamma(p) \Big(\int_\Omega |\nabla \mathbf{u}(\mathbf{x})|^2 \, dx + \int_\Omega |\mathbf{u}(\mathbf{x})|^2 \, dx \Big) \geq \gamma(p) \|\mathbf{u}\|^2_{H^1_0(\Omega)}.$$

This proves the coerciveness of $L(p)$. □

With the purpose of investigating stability properties of the solution $\mathbf{u}_L(\mathbf{x}, p)$ to (4.6.6)–(4.6.7), it is convenient to consider the representation of $\mathbf{u}_L$ in terms of the Green's function $\mathbf{H}(\mathbf{x}, \mathbf{x}'; p)$, that is, the solution to the problem

$$p^2 \mathbf{H}(\mathbf{x}, \mathbf{x}'; p) - \nabla' \cdot \{[\mathbf{G}_0(\mathbf{x}') + \mathbf{G}'_L(\mathbf{x}', p)] \nabla' \mathbf{H}(\mathbf{x}, \mathbf{x}'; p)\} = \delta(\mathbf{x} - \mathbf{x}') \mathbf{1}, \tag{4.6.8}$$

$$\mathbf{H}(\mathbf{x}, \mathbf{x}'; p) = 0, \qquad \mathbf{x}' \in \partial\Omega \tag{4.6.9}$$

where $\nabla' = \partial / \partial \mathbf{x}'$. The representation is

$$\mathbf{u}_L(\mathbf{x}, p) = \int_\Omega \mathbf{H}(\mathbf{x}, \mathbf{x}'; p) \, \mathbf{B}_L(\mathbf{x}', p) \, dx. \tag{4.6.10}$$

Concerning $\mathbf{H}(\mathbf{x}, \mathbf{x}'; p)$ we have the following lemma.[25]

LEMMA 4.6.3. *If* $\mathbf{G}$ *satisfies the conditions* (4.6.1)–(4.6.2) *then there exists a unique solution* $\mathbf{H}(\mathbf{x}, \mathbf{x}'; p)$, *to the problem* (4.6.8)–(4.6.9), *such that*

[25] Throughout, when p is complex-valued the writing $p \to \infty$ stands for $|p| \to \infty$.

$\mathbf{H}(\mathbf{x},\cdot;p) \in H_0^1(\Omega)$; $\mathbf{H}(\mathbf{x},\mathbf{x}';\cdot)$ *is continuous;* $\lim_{p\to\infty} p^{2-\varepsilon}\mathbf{H}(\mathbf{x},\mathbf{x}';p) = 0$ *almost everywhere in* Ω, $\varepsilon > 0$.

Proof. Since $L(p)$ is a coercive operator, by Lax-Milgram's theorem it follows that $L(p)$ is an isomorphism of $H_0^1(\Omega)$ onto $H^{-1}(\Omega)$. Then, since $\delta \in H^{-1}(\Omega)$, the existence and uniqueness of $\mathbf{H}(\mathbf{x},\cdot;p) \in H_0^1(\Omega)$ are proved.

Because $L(p)$ is a continuous function of p, by [120, Lemma 44.1], we can say that $\mathbf{H}(\mathbf{x},\mathbf{x}';\cdot)$ is a continuous function and hence is bounded in every bounded set.

To determine the behaviour of $\mathbf{H}(\mathbf{x},\mathbf{x}';\cdot)$ at infinity, observe that by (4.6.8)

$$\int_\Omega \mathbf{H}(\mathbf{x},\mathbf{x}';p)\phi(\mathbf{x}')\,dx' = \frac{1}{p^2}\left\{\int_\Omega \nabla'\cdot[(\mathbf{G}_0(\mathbf{x}') + \mathbf{G}_L'(\mathbf{x}',p))\nabla'\mathbf{H}(\mathbf{x},\mathbf{x}';p)]\phi(\mathbf{x}')\,dx' + \phi(\mathbf{x})\right\}$$

for every real-valued $\phi \in C_0^\infty(\Omega)$. Application of the divergence theorem gives

$$\int_\Omega \mathbf{H}(\mathbf{x},\mathbf{x}';p)\left\{\phi(\mathbf{x}') - \frac{1}{p^2}\nabla\cdot[(\mathbf{G}_0(\mathbf{x}') + \mathbf{G}_L'(\mathbf{x}',p))\nabla'\phi(\mathbf{x}')]\right\}dx' = \frac{1}{p^2}\phi(\mathbf{x}).$$

Since

$$\lim_{p\to\infty} p^2\int_\Omega \mathbf{H}(\mathbf{x},\mathbf{x}';p)\phi(\mathbf{x}')\,dx' = \int_\Omega \mathbf{H}(\mathbf{x},\mathbf{x}';p)\nabla\cdot[\mathbf{G}_0(\mathbf{x}')\nabla'\phi(\mathbf{x}')]\,dx' + \phi(\mathbf{x})$$

then

$$\lim_{p\to\infty} p^{2-\varepsilon}\int_\Omega \mathbf{H}(\mathbf{x},\mathbf{x}';p)\phi(\mathbf{x}')\,dx' = 0.$$

The arbitrariness of $\phi \in C_0^\infty(\Omega)$ provides the desired result. □

Incidentally, the behaviour of $\mathbf{H}$ as $p \to \infty$ allows us to write

$$\lim_{p\to\infty}\int_\Omega p^2\mathbf{H}(\mathbf{x},\mathbf{x}';p)\phi(\mathbf{x}')\,dx' = \phi(\mathbf{x}). \tag{4.6.11}$$

We are now in a position to prove the stability result we are looking for.

THEOREM 4.6.1. *If* $\mathbf{b}$ *and* $\mathbf{u}_0$ *satisfy* (4.6.3)–(4.6.5) *then there exists a unique solution*

$$\mathbf{u}(\mathbf{x},t) \in H^1(\mathbb{R}^+;L^2(\Omega)) \cap L^2(\mathbb{R}^+;H_0^1(\Omega))$$

to the problem (4.1.1)–(4.1.4) *such that*

$$\int_0^\infty\int_\Omega [|\nabla\mathbf{u}(\mathbf{x},t)|^2 + |\dot{\mathbf{u}}(\mathbf{x},t)|^2]dx < \infty.$$

Proof. For $p \in \mathbb{C}^+$, by (4.6.6) we have

$$\begin{aligned}\lim_{p\to\infty} \mathbf{B}_L(\mathbf{x},p) &= \lim_{p\to\infty} [\mathbf{b}_L(\mathbf{x},p) + p\,\mathbf{u}(\mathbf{x},0) + \dot{\mathbf{u}}(\mathbf{x},0) + \mathbf{U}_L(\mathbf{x},p)] \\ &= \lim_{p\to\infty} [p\,\mathbf{u}(\mathbf{x},0) + \dot{\mathbf{u}}(\mathbf{x},0)]\end{aligned}$$

Then, for every $\varepsilon > 0$, the representation (4.6.10) of $\mathbf{u}_L$ in terms of $\mathbf{H}$ and the asymptotic behaviour of $\mathbf{H}$ yield

$$\lim_{p\to\infty} p^{1-\varepsilon}\mathbf{u}_L(\mathbf{x},p) = \lim_{p\to\infty} p^{1-\varepsilon} \int_\Omega \mathbf{H}(\mathbf{x},\mathbf{x}';p)[p\mathbf{u}(\mathbf{x}',0) + \dot{\mathbf{u}}(\mathbf{x}',0)]dx' = 0.$$

Moreover, (4.6.11) gives

$$\begin{aligned}&\lim_{p\to\infty} p\,\mathbf{u}_L(\mathbf{x},p) \\ &= \lim_{p\to\infty} \left[\int_\Omega p^2\mathbf{H}(\mathbf{x},\mathbf{x}';p)\mathbf{u}(\mathbf{x}',0)\,dx' + p\int_\Omega \mathbf{H}(\mathbf{x},\mathbf{x}';p)\dot{\mathbf{u}}(\mathbf{x}',0)\,dx'\right] = \mathbf{u}(\mathbf{x},0).\end{aligned}$$

Hence we have

$$\lim_{p\to\infty} p^{1-\varepsilon}[p\mathbf{u}_L(\mathbf{x},p) - \mathbf{u}(\mathbf{x},0)] = 0.$$

Letting $p = i\beta$, $\beta \in \mathbb{R}$, we can regard $\mathbf{u}_F(\mathbf{x},p)$ and $\beta\mathbf{u}_F(\mathbf{x},\beta) - \mathbf{u}(\mathbf{x},0)$ as L^2-functions with respect to β. Indeed, they are the L^2-Fourier transforms of the function

$$\tilde{\mathbf{u}}(\mathbf{x},t) = \begin{cases} \mathbf{u}(\mathbf{x},t), & \text{as } t \geq 0, \\ 0, & \text{as } t < 0, \end{cases}$$

and of its first-order time derivative. Then Plancherel's theorem gives

$$\infty > \frac{1}{2\pi}\int_{-\infty}^{\infty} |\mathbf{u}_F(\mathbf{x},\beta)|^2 d\beta = \int_{-\infty}^{\infty} |\tilde{\mathbf{u}}(\mathbf{x},t)|^2 dt = \int_0^\infty |\mathbf{u}(\mathbf{x},t)|^2 dt,$$

$$\infty > \frac{1}{2\pi}\int_{-\infty}^{\infty} |i\beta\mathbf{u}_F(\mathbf{x},\beta) + \mathbf{u}(\mathbf{x},0)|^2\, d\beta = \int_0^\infty |\dot{\mathbf{u}}(\mathbf{x},t)|^2\, dt$$

almost everywhere in Ω.

Now we need an estimate for $\nabla\mathbf{u}_L$. We have

$$\nabla\mathbf{u}_L(\mathbf{x},p) = \int_\Omega \nabla\mathbf{H}(\mathbf{x},\mathbf{x}';p)\mathbf{b}_L(\mathbf{x}',p)\,dx';$$

of course, $\nabla\mathbf{H}(\mathbf{x},\mathbf{x}';p)$ is a solution to

$$p^2\,\nabla\mathbf{H}(\mathbf{x},\mathbf{x}';p) - \nabla'\cdot\{[\mathbf{G}_0(\mathbf{x}') + \mathbf{G}_L'(\mathbf{x}',p)]\nabla'\nabla\mathbf{H}(\mathbf{x},\mathbf{x}';p)\} = \nabla\delta(\mathbf{x}-\mathbf{x}')\mathbf{1}.$$

By paralleling Lemma 4.6.3 (cf. also [120], Lemma 23.2), we can prove that

$$\nabla \mathbf{H}(\mathbf{x}, \cdot; p) \in L^2(\Omega),$$

$$\nabla \mathbf{H}(\mathbf{x}, \mathbf{x}'; \cdot) \text{ is a continuous function on } \mathbb{C}^+,$$

$$\lim_{p\to\infty} p^{2-\varepsilon} \int_\Omega \nabla \mathbf{H}(\mathbf{x}, \mathbf{x}'; p)\phi(\mathbf{x}')\, dx' = 0, \qquad \phi \in C_0^\infty(\Omega),\ \varepsilon > 0.$$

Then Plancherel's theorem yields

$$\infty > \frac{1}{2\pi} \int_{-\infty}^{\infty} |\nabla \mathbf{u}_F(\mathbf{x}, \beta)|^2\, d\beta = \int_0^\infty |\nabla \mathbf{u}(\mathbf{x}, t)|^2\, dt$$

almost everywhere in Ω. In conclusion, we have

$$\int_0^\infty \int_\Omega [|\nabla \mathbf{u}(\mathbf{x}, t)|^2 + |\dot{\mathbf{u}}(\mathbf{x}, t)|^2]\, dx\, dt < \infty,$$

which completes the proof. □

The last step of this analysis consists in showing which properties hold for the solution $\mathbf{u}$ to (4.1.1)–(4.1.4). In this regard, letting $H(\Omega)$ be a Hilbert space, consider the spaces

$$L^2_{loc}(\mathbb{R}^+; H(\Omega)) = \{\mathbf{u} : \mathbb{R}^+ \to H(\Omega);\ \|\mathbf{u}(t)\|_H \in L^2_{loc}(\mathbb{R}^+)\},$$

$$L^2_l(\mathbb{R}^+; H(\Omega)) = \{\mathbf{u} \in L^2_{loc}(\mathbb{R}^+; H(\Omega));\ \|\mathbf{u}_L(p)\|_H < \infty \forall\ p \in \mathbb{C}^{++}\},$$

$$H^1_l(\mathbb{R}^+; H(\Omega)) = \{\mathbf{u}, \dot{\mathbf{u}} \in L^2_l(\mathbb{R}^+; H(\Omega))\}.$$

THEOREM 4.6.2. *If* $\mathbf{b} \in L^2_l(\mathbb{R}^+; H^{-1}(\Omega))$, $\mathbf{u}_0(\mathbf{x}, t)$ *satisfies* (4.6.4), *and* $\mathbf{U} \in L^2_l(\mathbb{R}^+;\ H^{-1}(\Omega))$, *then there exists a unique solution* $\mathbf{u}$ *to the problem* (4.1.1)–(4.1.4) *such that*

$$\mathbf{u} \in H^1_l(\mathbb{R}^+; L^2(\Omega)) \cap L^2_l(\mathbb{R}^+; H^1_0(\Omega)).$$

Proof. The Laplace transform $\mathbf{B}_L(\mathbf{x}, p)$ is well defined for every $p \in \mathbb{C}^{++}$. Then there exist the inverse Laplace transforms of $\mathbf{u}_L(\mathbf{x}, p)$ and $p\mathbf{u}_L(\mathbf{x}, p) - \mathbf{u}(\mathbf{x}, 0)$ for every $p \in \mathbb{C}^{++}$, that is, $\mathbf{u}(\mathbf{x}, t),\ \dot{\mathbf{u}}(\mathbf{x}, t) \in L^2_l(\mathbb{R}^+; L^2(\Omega))$. By (4.6.10),

$$\nabla \mathbf{u}_L(\mathbf{x}, p) = \int_\Omega \nabla \mathbf{H}(\mathbf{x}, \mathbf{x}'; p)\, \mathbf{B}_L(\mathbf{x}', p)\, dx'.$$

The properties of $\nabla \mathbf{H}$, shown in Theorem 4.6.1, imply that there exists the inverse Laplace transform of $\nabla \mathbf{u}(\mathbf{x}, p)$ for every $p \in \mathbb{C}^{++}$, which means that $\mathbf{u}(\mathbf{x}, t) \in L^2_l(\mathbb{R}^+; H^1_0(\Omega))$. □

4.7. Asymptotic behaviour: exponential decay. Concerning the properties of the solution to the problem (4.1.1)–(4.1.4), it is of interest to investigate the asymptotic behaviour in terms of Liapunov functionals [50]. In this regard we recall that, as shown in §3.5, two forms are possible for the free energy of a linear viscoelastic material. Here we have recourse to the Volterra type free energy and show that it is a profitable Liapunov functional for the problem (4.1.1)–(4.1.4) [68]. First, however, we make it clear that the investigation is performed by assuming that the relaxation function $\mathbf{G}$ satisfies suitable restrictions. These are as follows. For any $\mathbf{v} \in C_0^\infty(\Omega)$,

$$-\int_\Omega \nabla\mathbf{v}(\mathbf{x}) \cdot \mathbf{G}'(\mathbf{x}, s)\nabla\mathbf{v}(\mathbf{x})\, dx > 0, \tag{4.7.1}$$

$$\int_\Omega \nabla\mathbf{v}(\mathbf{x}) \cdot \mathbf{G}''(\mathbf{x}, s)\nabla\mathbf{v}(\mathbf{x})\, dx > 0. \tag{4.7.2}$$

Further, there exists a positive constant κ such that

$$\int_\Omega \nabla\mathbf{v}(\mathbf{x}) \cdot [\mathbf{G}''(\mathbf{x}, s) + \kappa\mathbf{G}'(\mathbf{x}, s)]\nabla\mathbf{v}(\mathbf{x})\, dx > 0. \tag{4.7.3}$$

Finally,

$$\int_0^\infty |\mathbf{G}_\infty(\mathbf{x}) - \mathbf{G}(\mathbf{x}, t)|\, dt < \infty. \tag{4.7.4}$$

The condition (4.7.4) ensures that initial histories $\mathbf{u}_0$ with constant gradient $\nabla\mathbf{u}(\mathbf{x}, \cdot)$ in time are admissible. As a whole, the conditions (4.7.1)–(4.7.4) are not severely restrictive in that they are satisfied by a large class of relaxation functions. In particular, relaxation functions of exponential type belong to this class.

Let $\hat{\mathbf{u}}^t$ be the difference history defined by

$$\hat{\mathbf{u}}^t(\mathbf{x}, s) = \mathbf{u}^t(\mathbf{x}, s) - \mathbf{u}(\mathbf{x}, t)$$

and consider the free energy functional

$$\begin{aligned}\Psi(t) = \tfrac{1}{2}\int_\Omega \Big\{&\nabla\mathbf{u}(\mathbf{x}, t) \cdot \mathbf{G}_\infty(\mathbf{x})\nabla\mathbf{u}(\mathbf{x}, t) \\ &- \int_0^\infty \nabla\hat{\mathbf{u}}^t(\mathbf{x}, s) \cdot \mathbf{G}'(\mathbf{x}, s)\nabla\hat{\mathbf{u}}^t(\mathbf{x}, s)\, ds\Big\}dx.\end{aligned} \tag{4.7.5}$$

Then we define the total energy of the body E as

$$E(t) = \frac{1}{2}\int_\Omega |\dot{\mathbf{u}}(\mathbf{x},t)|^2 dx + \Psi(t). \tag{4.7.6}$$

In terms of E we prove shortly the main result of this section. Preliminarily, though, it is convenient to write the problem (4.1.1)–(4.1.4) in the form

$$\frac{\partial}{\partial t}\mathbf{u}(\mathbf{x},t) = \mathbf{v}(\mathbf{x},t), \qquad \mathbf{u}(\mathbf{x},0) = \mathbf{u}_0(\mathbf{x},0), \tag{4.7.7}$$

$$\begin{gathered}\frac{\partial}{\partial t}\mathbf{v}(\mathbf{x},t) = \nabla\cdot\left[\mathbf{G}_\infty(\mathbf{x})\nabla\mathbf{u}(\mathbf{x},t) + \int_0^\infty \mathbf{G}'(\mathbf{x},s)\nabla\hat{\mathbf{u}}^t(\mathbf{x},s)\,ds\right],\\ \mathbf{v}(\mathbf{x},0) = \frac{\partial\mathbf{u}_0}{\partial t}(\mathbf{x},0),\end{gathered} \tag{4.7.8}$$

$$\begin{gathered}\frac{\partial}{\partial t}\hat{\mathbf{u}}^t(\mathbf{x},s) = -\frac{\partial}{\partial s}\hat{\mathbf{u}}^t(\mathbf{x},s) - \mathbf{v}(\mathbf{x},t), \qquad \hat{\mathbf{u}}^0(\mathbf{x},s) = \mathbf{u}_0(\mathbf{x},-s) - \mathbf{u}_0(\mathbf{x},0),\\ s\in\mathbb{R}^+.\end{gathered} \tag{4.7.9}$$

Equations (4.7.7)–(4.7.9) are in a form which is especially suited for the application of the theory of contraction semigroups in Hilbert spaces. For brevity let χ be the triplet

$$\chi(\mathbf{x},t) = (\mathbf{u}(\mathbf{x},t),\ \mathbf{v}(\mathbf{x},t),\ \hat{\mathbf{u}}^t(\mathbf{x},\cdot))$$

and let $\mathcal{H}$ be the Hilbert space of all triplets χ such that $\mathbf{u}(\cdot,t)\in H_0^1(\Omega)$, $\mathbf{v}(\cdot,t)\in L^2(\Omega)$ almost everywhere on $\mathbb{R}^+$ and

$$-\int_0^\infty\int_\Omega \nabla\hat{\mathbf{u}}^t(\mathbf{x},s)\cdot\mathbf{G}'(\mathbf{x},s)\nabla\hat{\mathbf{u}}^t(\mathbf{x},s)\,dx\,ds < \infty.$$

The inner product of two elements $\chi_1,\chi_2\in\mathcal{H}$ is taken as

$$\begin{aligned}(\chi_1,\chi_2) = \int_\Omega\Bigg[&\nabla\mathbf{u}_1(\mathbf{x},t)\cdot\mathbf{G}_\infty(\mathbf{x})\nabla\mathbf{u}_2(\mathbf{x},t) + \mathbf{v}_1(\mathbf{x},t)\cdot\mathbf{v}_2(\mathbf{x},t)\\ &-\int_0^\infty \nabla\hat{\mathbf{u}}_1^t(\mathbf{x},s)\cdot\mathbf{G}'(\mathbf{x},s)\nabla\hat{\mathbf{u}}_2^t(\mathbf{x},s)\,ds\Bigg]dx\end{aligned}$$

and, of course, the norm is taken as the energy norm

$$\|\chi\| = (\chi,\chi)^{1/2}.$$

Now we prove some preliminary results about the system (4.7.7)–(4.7.9) [50].

LEMMA 4.7.1. *If the body force* $\mathbf{b}$ *is zero then the energy* (4.7.6) *is an integrable function on* $\mathbb{R}^+$.

Proof. By (4.7.5), (4.7.6), and the assumption (4.7.2) we have

$$\frac{d}{dt}E(t) = -\frac{1}{2}\int_\Omega \int_0^\infty \nabla\hat{\mathbf{u}}^t(s)\cdot \mathbf{G}''(\mathbf{x},s)\nabla\hat{\mathbf{u}}^t(s)\,ds\,dx \le 0 \tag{4.7.10}$$

whence

$$0 \le E(t) \le E(0), \qquad t\in\mathbb{R}^+.$$

Integration of (4.7.10) on $\mathbb{R}^+$ yields

$$\int_0^\infty \int_\Omega \int_0^\infty \nabla\hat{\mathbf{u}}^t(\mathbf{x},s)\cdot \mathbf{G}''(\mathbf{x},s)\nabla\hat{\mathbf{u}}^t(\mathbf{x},s)\,ds\,dx\,dt$$
$$= 2\lim_{t\to\infty}[E(0)-E(t)] \le 2E(0),$$

which in turn, along with (4.7.3), yields

$$-\int_0^\infty \int_\Omega \int_0^\infty \nabla\hat{\mathbf{u}}^t(\mathbf{x},s)\cdot \mathbf{G}'(\mathbf{x},s)\nabla\hat{\mathbf{u}}^t(\mathbf{x},s)\,ds\,dx\,dt < \infty.$$

Then, because of Theorem 4.6.1, we have

$$\int_0^\infty E(t)dt = \int_0^\infty \int_\Omega \Bigg[\nabla\mathbf{u}(\mathbf{x},t)\cdot\mathbf{G}_\infty\nabla\mathbf{u}(\mathbf{x},t) + |\dot{\mathbf{u}}(\mathbf{x},t)|^2$$
$$- \int_0^\infty \nabla\hat{\mathbf{u}}^t(\mathbf{x},s)\cdot\mathbf{G}'(\mathbf{x},s)\nabla\hat{\mathbf{u}}^t(\mathbf{x},s)ds\Bigg]\,dx\,dt < \infty,$$

which completes the proof. □

Observe that the system (4.7.7)–(4.7.9) can be written as

$$\dot{\chi} = A\chi, \qquad \chi(0) = \chi_0$$

where A stands for the operator defined by the right-hand side of the three equations. The domain $\mathcal{D}(A)$ of A is defined as

$$\mathcal{D}(A) = \Big\{\chi\in\mathcal{H};\ \frac{\partial}{\partial s}\hat{\mathbf{u}}^t(\cdot,s)\in H_0^1(\Omega);$$
$$\int_0^\infty\int_\Omega \frac{\partial}{\partial s}\nabla\hat{\mathbf{u}}(\mathbf{x},t-s)\cdot\mathbf{G}'(\mathbf{x},s)\frac{\partial}{\partial s}\nabla\hat{\mathbf{u}}(\mathbf{x},t-s)\,dx\,ds<\infty;$$
$$\nabla\cdot\Big[\mathbf{G}_\infty(\cdot)\nabla\hat{\mathbf{u}}(\cdot,t) + \int_0^\infty \mathbf{G}'(\cdot,s)\nabla\hat{\mathbf{u}}(\cdot,t-s)\,ds\Big]\in H^1(\Omega)\Big\}.$$

LEMMA 4.7.2. *If* $\mathbf{G}$ *satisfies* (4.7.4) *then any history* $\mathbf{u}^t$ *such that* $\chi(\mathbf{x},t) = (\mathbf{u}(\mathbf{x},t), \dot{\mathbf{u}}(\mathbf{x},t),\ \mathbf{u}^t(\mathbf{x},s) - \mathbf{u}(\mathbf{x},t)) \in \mathcal{D}(A)$ *satisfies* (4.6.4)–(4.6.5) *and* $A\chi \in \mathcal{H}$.

Proof. By the definition of $\mathcal{H}$, every $\chi \in \mathcal{H}$ meets the condition (4.6.4). Letting $\phi \in H_0^1(\Omega)$, by the Cauchy-Schwarz inequality we have

$$\int_0^\infty \left| \int_\Omega \int_t^\infty -\nabla \mathbf{u}^t(\mathbf{x},s) \cdot \mathbf{G}'(\mathbf{x},s) \nabla \phi(\mathbf{x})\, ds\, dx \right|^2 dt$$

$$\leq \int_0^\infty \left[\int_\Omega \int_0^\infty -\nabla \mathbf{u}^t(\mathbf{x},s) \cdot \mathbf{G}'(\mathbf{x},s) \nabla \mathbf{u}^t(\mathbf{x},s) ds\, dx \right.$$

$$\left. \int_\Omega \int_t^\infty -\nabla \phi(\mathbf{x}) \cdot \mathbf{G}'(\mathbf{x},s) \nabla \phi(\mathbf{x}) ds\, dx \right] dt$$

$$= \int_0^\infty \left[\int_\Omega \int_0^\infty -\nabla \mathbf{u}^t(\mathbf{x},s) \cdot \mathbf{G}'(\mathbf{x},s) \nabla \mathbf{u}^t(\mathbf{x},s) ds\, dx \right.$$

$$\left. \int_\Omega \nabla \phi(\mathbf{x}) \cdot [\mathbf{G}(\mathbf{x},t) - \mathbf{G}_\infty(\mathbf{x})] \nabla \phi(\mathbf{x}) dx \right] dt.$$

By the definition of $\mathcal{H}$ the integral on $\Omega \times \mathbb{R}^+$ is finite for every $t \in \mathbb{R}^+$. Then by the condition (4.7.4) it follows that the right-hand side is finite, whence we have (4.6.5). □

Since

$$(A\chi, \chi) = -\frac{1}{2} \int_0^\infty \int_\Omega \nabla \hat{\mathbf{u}}^t(\mathbf{x},s) \cdot \mathbf{G}''(\mathbf{x},s) \nabla \hat{\mathbf{u}}^t(\mathbf{x},s)\, ds \leq 0,$$

following Dafermos [34] we state the following properties of A.

LEMMA 4.7.3. *The operator* A *is dissipative, namely* $(A\chi, \chi) \leq 0,\ \chi \in \mathcal{D}(A)$, *and the range of* $A - I$ *is* $\mathcal{H}$.

The properties established by Lemma 4.7.3 allow the application of the Lumer-Phillips theorem whereby A generates a strongly continuous semigroup of linear contractions $S(t)$ on $\mathcal{H}$ relative to the norm $\|\chi\|$. Then we can write the solution to (4.7.7)–(4.7.9) as

$$\chi(t) = S(t)\, \chi_0.$$

Since

$$E(t) = \frac{1}{2}(S(t)\, \chi_0,\ S(t)\, \chi_0) \tag{4.7.11}$$

then by Lemma 4.7.1 we have

$$\int_0^\infty (S(t)\, \chi_0,\ S(t)\, \chi_o)\, dt < \infty,$$

which holds for every $\chi_0 \in \mathcal{H}$ in that $\mathcal{D}(A)$ is dense in $\mathcal{H}$.

This allows the application of Datko-Pazy's theorem whereby for a strongly continuous semigroup of linear operators $S(t)$ on a Hilbert space $\mathcal{H}$ such that

$$\int_0^\infty (S(t)\,\chi_0,\ S(t)\,\chi_0)dt < \infty, \qquad \chi_0 \in \mathcal{H},$$

there exist two positive constants M, μ such that

$$(S(t)\,\chi_0,\ S(t)\,\chi_0) \le M \exp(-\mu t)(\chi_0, \chi_0). \tag{4.7.12}$$

Remark 4.7.1. Quite often what we call Datko-Pazy's theorem is quoted as Pazy's theorem [107]. Really, by extending a well-known theorem of Liapunov concerning Hurwitzian matrices, Datko [36] proved the following theorem.

A necessary and sufficient condition that a strongly continuous semigroup of operators $S(t)$ of class C^0 defined on a complex Hilbert space X satisfy the condition $\|S(t)\| \le M \exp(-\mu t)$, with $M \in [1, \infty)$, $\mu \in \mathbb{R}^{++}$, is the existence of an Hermitian endomorphism B on X, with $B \ge 0$, such that for all x in the domain of the infinitesimal generator A of $S(t)$ the relation $2(BAx, x) = -\|x\|^2$ holds.

He also gave the following corollary.

A necessary and sufficient condition that a strongly continuous semigroup $S(t)$ of class C^0 defined on a complex Hilbert space satisfy the inequality $\|S(t)\| \le M \exp(-\mu t)$ is that for each $x \in X$ the integral $\int_0^\infty \|S(t)\|^2 dt$ be convergent.

Later, while investigating Hermitian forms in X such that $(Bx, x) \ge b\|x\|^2$, Pazy [107] proved the following theorem.

Let $p \ge 1$ be fixed, X a Banach space, $S(T)$ a strongly continuous semigroup on X. The norm $\|x\|_p = (\int_0^\infty \|S(t)x\|^p dt)^{1/p}$ is finite for every $x \in X$ if and only if there exist constants $M \in [1, \infty)$ and $\mu \in \mathbb{R}^{++}$ such that $\|S(t)\| \le M \exp(-\mu t)$.

So we have applied Datko's corollary or Pazy's theorem in the case $p = 2$.

We are now in a position to prove the desired result on the asymptotic behaviour of the solution [50].

THEOREM 4.7.1. *If the body force* **b** *is zero then, for any solution* **u** *to the problem* (4.1.1)–(4.1.4), *there exist two positive constants* M, μ *such that*

$$E(t) \le M\ \exp(-\mu t)E(0). \tag{4.7.13}$$

Proof. Because $S(0) = I$, by (4.7.11) and (4.7.12) we find the sought result (4.7.13). $\square$

4.8. Existence, uniqueness, and stability for fluids. Cauchy's problem is now investigated for the viscoelastic fluid as modelled in §3.6. As usual, we confine the attention to incompressible fluids (with $\rho = 1$) and to the linear approximation of the equation of motion.

Let $\Omega \subset \mathcal{E}^3$ be a smooth domain occupied by the linear viscoelastic fluid and let $\mathbf{v} : \Omega \to V$ be the velocity field. Cauchy's problem consists of finding the fields $\mathbf{v}$ and Π on $\Omega \times \mathbb{R}^+$ such that

$$\nabla \cdot \mathbf{v}(\mathbf{x}, t) = 0 \quad \text{in } \Omega \times \mathbb{R}^{++}, \tag{4.8.1}$$

$$\dot{\mathbf{v}}(\mathbf{x}, t) = -\nabla \Pi(\mathbf{x}, t) + \int_0^\infty \mu(s)\, \Delta \mathbf{v}(\mathbf{x}, t - s)\, ds + \mathbf{b}(\mathbf{x}, t) \quad \text{in } \Omega \times \mathbb{R}^+, \tag{4.8.2}$$

$$\mathbf{v}(\mathbf{x}, t) = 0 \quad \text{on } \partial\Omega \times \mathbb{R}^+, \tag{4.8.3}$$

$$\mathbf{v}(\mathbf{x}, t) = \mathbf{v}_0(\mathbf{x}, t) \quad \text{in } \Omega \times \mathbb{R}^-. \tag{4.8.4}$$

Cauchy's problem (4.8.1)–(4.8.4) has been investigated in several papers. Among others, we mention those by Craik [30], Joseph [85], [86], and Slemrod [116], [117]. The analogous problem for a finite linear constitutive equation was examined, e.g., by Infante and Walker [84]. For local and global existence results in nonlinear viscoelasticity, we mention [112] also in connection with a relevant review of the literature on the subject.

Concerning linear viscoelastic fluids, Slemrod [116] proved that if, in our notation,

$$\mu \in C^2(\mathbb{R}^+), \qquad \mu(s) \to 0 \quad \text{as } s \to \infty$$

and

$$\mu(s) > 0, \qquad \mu'(s) < 0, \qquad \mu''(s) \geq 0, \qquad s \in \mathbb{R}^+,$$

then the rest state of the fluid is stable, in the sense of a norm of "fading memory" type, and the solution to (4.8.1)–(4.8.4) converges to the rest state in this norm as $t \to \infty$. Next he proved that the additional assumption

$$-\int_0^\infty \mu'(s)\, s^2\, ds < \infty$$

yields asymptotic stability. Later [117] he showed that the further assumption

$$\mu'(s) + \xi\, \mu(s) \leq 0, \qquad \mu''(s) + \xi\, \mu'(s) \geq 0, \qquad s \in \mathbb{R}^+,$$

for some $\xi \in \mathbb{R}^{++}$, implies exponential decay in an appropriate fading memory norm.

Here different assumptions on the constitutive properties are made. The relaxation function μ is required to satisfy the principle of fading memory in the form

$$\text{(4.8.5)} \qquad \mu \in C(\mathbb{R}^+, \mathbb{R}), \qquad \int_0^\infty |\mu(s)|ds < \infty,$$

and the second law of thermodynamics, viz.

$$\text{(3.7.5)} \qquad \mu_c(\omega) > 0, \qquad \omega \in \mathbb{R}.$$

As we show shortly, the two conditions (4.8.5), (3.7.5) alone are sufficient to enforce existence, uniqueness, and stability of the solution to the problem (4.8.1)–(4.8.4).[26] This allows us to view (3.7.5) as the right physical condition on the shear relaxation function μ instead of the convexity condition.

For later convenience we introduce the following notation. Let $\overset{\sigma}{L}{}^2(\Omega)$ be the Hilbert space obtained by the completion of solenoidal vector fields $\mathbf{v}(\mathbf{x}) \in C_0^\infty(\Omega)$ in the $L^2(\Omega)$ inner product. Similarly, let $\overset{\sigma}{H}{}_0^1(\Omega)$ be the Hilbert space obtained by the completion of solenoidal vector fields in the $H_0^1(\Omega)$-norm. Moreover, let $\overset{\pi}{L}{}^2(\Omega)$ be the Hilbert space obtained by the completion of irrotational vector fields $\mathbf{v}(\mathbf{x}) \in C_0^\infty(\Omega)$ in the $L^2(\Omega)$-norm. Then we have $L^2(\Omega) = \overset{\sigma}{L}{}^2(\Omega) \oplus \overset{\pi}{L}{}^2(\Omega)$. The symbol $\overset{\sigma}{H}{}^{-1}(\Omega)$ denotes the dual of $\overset{\sigma}{H}{}_0^1(\Omega)$.

Letting, as usual, a subscript L denote the Laplace transform (in time), we write the problem (4.8.1)–(4.8.4) in the form

$$\text{(4.8.6)} \qquad \nabla \cdot \mathbf{v}_L(\mathbf{x}, p) = 0 \quad \text{in } \Omega,$$

$$\text{(4.8.7)} \qquad p\,\mathbf{v}_L(\mathbf{x}, p) = -\nabla \Pi_L(\mathbf{x}, p) + \mu(p)\,\Delta \mathbf{v}_L(\mathbf{x}, p) + \tilde{\mathbf{b}}(\mathbf{x}, p) \quad \text{in } \Omega,$$

$$\text{(4.8.8)} \qquad \mathbf{v}_L(\mathbf{x}, p) = 0 \quad \text{on } \partial\Omega$$

where

$$\tilde{\mathbf{b}}(\mathbf{x}, p) = \mathbf{b}_L(\mathbf{x}, p) + \mathbf{v}_0(\mathbf{x}, 0) + \int_0^\infty \exp(-ps) \int_s^\infty \mu(\tau)\,\Delta \mathbf{v}_0(\mathbf{x}, s - \tau)\,d\tau\,ds.$$

Observe that the hypotheses $\mathbf{b} \in L^2(\mathbb{R}^+; H^{-1}(\Omega))$, $\mathbf{v}_0 \in \overset{\sigma}{H}{}_0^1(\Omega)$, and

$$\mathbf{V}_0(\mathbf{x}, t) := \int_t^\infty \mu(s) \Delta \mathbf{v}_0(\mathbf{x}, t - s)\,ds \in L^2(\mathbb{R}^+; \overset{\sigma}{H}{}^{-1}(\Omega))$$

[26]Here we follow [51].

make $\tilde{\mathbf{b}}(\mathbf{x},p)$ well defined for any $p \in \mathbb{C}^+$.

Following [118], Lemma 2.1, we can say that if $\mathbf{v}_L$ is a solution to

$$\int_\Omega [\mu_L(p)\nabla \mathbf{v}_L(\mathbf{x},p) \cdot \nabla \mathbf{w}(\mathbf{x}) + p\,\mathbf{v}_L(\mathbf{x},p)\mathbf{w}(\mathbf{x})]dx$$

$$= \int_\Omega \tilde{\mathbf{b}}(\mathbf{x},p) \cdot \mathbf{w}(\mathbf{x})dx, \qquad \mathbf{w} \in \overset{\sigma}{H}{}^1_0(\Omega), \tag{4.8.9}$$

then there exists a scalar field in $L^2(\Omega)$, say Π_L, such that (4.8.6) and (4.8.7) hold in Ω in the distributional sense and, meanwhile, (4.8.8) holds. This motivates the following definition.

DEFINITION 4.8.1. A function $\mathbf{v}_L \in \overset{\sigma}{H}{}^1_0(\Omega)$ is called weak solution to (4.8.6)–(4.8.8) if (4.8.9) holds.

By general theorems on elliptic systems of equations [118], [120], [56], the coerciveness of the bilinear form

$$a(\mathbf{v},\mathbf{w};p) = \int_\Omega [\mu_L(p)\nabla \mathbf{v}(\mathbf{x}) \cdot \nabla \mathbf{w}^*(\mathbf{x}) + p\mathbf{v}(\mathbf{x}) \cdot \mathbf{w}^*(\mathbf{x})]dx$$

in $\overset{\sigma}{H}{}^1_0(\Omega)$ enforces the existence of a (weak) solution to (4.8.9) for every $\tilde{\mathbf{b}} \in \overset{\sigma}{H}{}^{-1}(\Omega)$. Then to prove the existence of a solution to (4.8.9), we have to show that there exists $\alpha \in \mathbb{R}^{++}$, possibly dependent on p, such that

$$|a(\mathbf{v},\mathbf{v};p)| \geq \alpha(p)\|\mathbf{v}\|^2_{\overset{\sigma}{H}{}^1_0(\Omega)}. \tag{4.8.10}$$

In fact, the coerciveness of a is a consequence of (3.7.5). To show that this is so we begin by recalling a property of Fourier integrals.

LEMMA 4.8.1. *Let $\phi,\psi \in L^1(\mathbb{R})$ with $\phi\psi \in L^1(\mathbb{R})$. If*

$$\Phi(\omega) = \int_{-\infty}^{\infty} \phi_F(\tau)\,\psi_F(\omega-\tau)\,d\tau$$

is continuous in $\mathbb{R}$ then

$$\int_{-\infty}^{\infty} \exp(-i\omega s)\,\phi(s)\,\psi(s) = \frac{1}{2\pi}\int_{-\infty}^{\infty} \phi_F(\tau)\,\psi_F(\omega-\tau)\,d\tau. \tag{4.8.11}$$

The proof is immediate and can be found, e.g., in [6], Chap. III.

LEMMA 4.8.2. *If (3.7.5) holds and $\mu \in L^1(\mathbb{R}^+)$ then*

$$\int_0^{\infty} \exp(-\alpha s)\,\cos\omega s\,\mu(s)\,ds > 0, \qquad \alpha \in \mathbb{R}^+, \qquad \omega \in \mathbb{R}. \tag{4.8.12}$$

Proof. Observe that when $\alpha = 0$ the inequality (4.8.12) coincides with (3.7.5) and then holds. Let $\alpha > 0$ and consider the functions

$$\phi(s) = \mu(|s|),\ s \in \mathbb{R}, \qquad \psi(s) = \begin{cases} 0, & s \in \mathbb{R}^{--}, \\ \exp(-\alpha s), & s \in \mathbb{R}^{+}. \end{cases}$$

It follows at once that $\phi, \psi, \phi\psi \in L^1(\mathbb{R})$ and

$$\phi_F(\tau) = 2\int_0^\infty \cos\tau s\ \mu(s)\,ds, \qquad \psi_F(\tau) = \frac{1}{\alpha + i\tau}.$$

Then

$$\Phi(\omega) = 2\int_{-\infty}^{\infty} \frac{1}{\alpha + i(\omega - \tau)} \int_0^\infty \cos\tau s\ \mu(s)\,ds\,d\tau$$

is in $C(\mathbb{R})$. Thus Lemma 4.8.1 applies and we have

$$\int_{-\infty}^{\infty} \exp[-(\alpha + i\omega)s]\,\mu(s)\,ds = \frac{1}{\pi}\int_0^\infty \frac{1}{\alpha + i(\omega - \tau)} \int_0^\infty \cos\tau s\ \mu(s)\,ds\,d\tau.$$

The real part yields

$$\int_0^\infty \exp(-\alpha s)\cos\omega s\ \mu(s)\,ds = \frac{1}{\pi}\int_{-\infty}^{\infty} \frac{\alpha}{\alpha^2 + (\omega - \tau)^2} \int_0^\infty \cos\tau s\ \mu(s)\,ds\,d\tau.$$

The condition (3.7.5) provides the desired result (4.8.12). □

We are now in a position to establish the coerciveness of the bilinear form a.

THEOREM 4.8.1. *If the relaxation function μ meets the conditions* (4.8.5) *and* (3.7.5), *then the bilinear form $a(\mathbf{v}, \mathbf{w}; p)$ is coercive for every $p \in \mathbb{C}^+$.*

Proof. Since $|a(\mathbf{v}, \mathbf{w}; p)| \geq \Re a(\mathbf{v}, \mathbf{w}; p)$, we need to show that

$$\Re a(\mathbf{v}, \mathbf{v}; p) \geq \nu(p)\|\mathbf{v}\|^2_{H^1_{0\sigma}(\Omega)}, \qquad \nu \in \mathbb{R}^{++}, \tag{4.8.13}$$

for any $p \in \mathbb{C}^+$. Letting $p = \alpha + i\beta, \alpha \in \mathbb{R}^+$, $\beta \in \mathbb{R}$, we have

$$\mu_L(p) = \int_0^\infty \exp(-\alpha s)\,\mu(s)\cos\beta s\,ds - i\int_0^\infty \exp(-\alpha s)\,\mu(s)\sin\beta s\,ds$$

whence

$$\Re a(\mathbf{v}, \mathbf{v}; p) = \int_0^\infty \exp(-\alpha s)\,\mu(s)\cos\beta s\,ds \int_\Omega |\nabla\mathbf{v}(\mathbf{x})|^2 dx + \int_\Omega |\mathbf{v}(\mathbf{x})|^2 dx.$$

Hence by Korn's inequality we obtain

$$\Re a(\mathbf{v}, \mathbf{v}; p) \geq c(\Omega)\Big[\int_0^\infty \exp(-\alpha s)\,\mu(s)\cos\beta s\,ds\Big]\ \|\mathbf{v}\|^2_{H^1_{0\sigma}(\Omega)}$$

where $c(\Omega)$ is a strictly positive constant which depends on the domain Ω. Recourse to Lemma 4.8.2, and then to the inequality (4.8.12), completes the proof. □

Now that by Theorem 4.8.1 we are guaranteed the existence and uniqueness of the solution to the problem (4.8.6)–(4.8.8), we can investigate the properties of the solution $\mathbf{v}_L$. To this end we consider a representation in terms of the Green's tensor function $\mathbf{\Gamma}$ such that

$$\begin{aligned}&\int_\Omega [\mu_L(p)\nabla'\mathbf{\Gamma}(\mathbf{x},\mathbf{x}';p)\nabla'\mathbf{v}(\mathbf{x}') + p\,\mathbf{\Gamma}(\mathbf{x},\mathbf{x}';p)\mathbf{v}(\mathbf{x}')]dx' \\ &\qquad = \int_\Omega \delta(\mathbf{x}-\mathbf{x}')\,\mathbf{v}(\mathbf{x}')dx', \qquad \mathbf{v} \in \overset{\sigma}{H}{}^1_0(\Omega),\end{aligned} \tag{4.8.14}$$

δ denoting Dirac's delta function. In terms of $\mathbf{\Gamma}$, the solution to (4.8.6)–(4.8.8) can be written as

$$\mathbf{v}_L(\mathbf{x},p) = \int_\Omega \mathbf{\Gamma}(\mathbf{x},\mathbf{x}';p)\tilde{\mathbf{b}}(\mathbf{x}',p)\,dx'. \tag{4.8.15}$$

The next results are based on the assumption that

$$\mathbf{b} \in L^2(\mathbb{R}^+; H^{-1}(\Omega)), \qquad \mathbf{v}_0 \in \overset{\sigma}{H}{}^1_0(\Omega), \tag{4.8.16}$$

and

$$\mathbf{V}_0 \in L^2(\mathbb{R}^+, \overset{\sigma}{L}{}^2(\Omega)).$$

First we prove existence, uniqueness, and behaviour of $\mathbf{\Gamma}$ with respect to p.

LEMMA 4.8.3. *If* (4.8.16) *holds then there exists a unique solution* $\mathbf{\Gamma}$ *to the problem* (4.8.14) *such that, for any* $p \in \mathbb{C}^+$,

(i) $\mathbf{\Gamma}(\mathbf{x},\cdot;p) \in \overset{\sigma}{H}{}^1_0(\Omega)$;
(ii) $\mathbf{\Gamma}(\mathbf{x},\mathbf{x}';\cdot)$ is continuous;
(iii) $\lim\ p^{1-\varepsilon}\mathbf{\Gamma}(\mathbf{x},\mathbf{x}';p) = 0$ as $p \to \infty$, $\varepsilon \in \mathbb{R}^{++}$.

Proof. The proof of (i) is a consequence of Theorem 4.8.1 and of Lax-Milgram's theorem because $\delta(\mathbf{x}-\mathbf{x}')$, $\mathbf{x} \in \Omega$, is in $\overset{\sigma}{H}{}^{-1}(\Omega)$. To prove (ii) observe that the bilinear form $a(\mathbf{v},\mathbf{w};\cdot)$ is continuous with respect to the third argument and then, by [120, Lemma 44.1], it follows the continuity of $\mathbf{\Gamma}(\mathbf{x},\mathbf{x}';\cdot)$. In regard to (iii), by (4.8.14) we have

$$\int_\Omega p^{1-\varepsilon}\,\mathbf{\Gamma}(\mathbf{x},\mathbf{x}';p)\left[\mathbf{v}(\mathbf{x}') - \frac{\mu_L(p)}{p}\Delta'\mathbf{v}(\mathbf{x}')\right]dx' = p^{-\alpha}\mathbf{v}(\mathbf{x})$$

for any $\varepsilon \in \mathbb{R}^+$ and $\mathbf{v} \in C_0^\infty(\Omega)$. Taking the limit as $p \to \infty$ yields (iii). □

Further recourse to Green's function provides a representation for $\nabla \mathbf{v}_L(\mathbf{x}, p)$. For convenience denote by $\nabla \mathbf{\Gamma}(\mathbf{x}, \mathbf{x}'; p)$ the (third-order) tensor function such that

$$\int_\Omega [\mu_L(p)\nabla' \, \nabla \mathbf{\Gamma}(\mathbf{x}, \mathbf{x}'; p)\nabla' \mathbf{w}(\mathbf{x}') + p(\nabla \mathbf{\Gamma}(\mathbf{x}, \mathbf{x}'; p))\mathbf{w}(\mathbf{x}')] \, dx'$$

$$= \int_\Omega [\nabla \delta(\mathbf{x} - \mathbf{x}')]\mathbf{w}(\mathbf{x}') \, dx' \tag{4.8.17}$$

for every $\mathbf{w} \in \overset{\sigma}{H}{}^1_0(\Omega)$. In terms of $\nabla \mathbf{\Gamma}$ we have

$$\nabla \mathbf{v}_L(\mathbf{x}; p) = \int_\Omega \nabla \mathbf{\Gamma}(\mathbf{x}, \mathbf{x}'; p) \, \tilde{\mathbf{b}}(\mathbf{x}'; p) \, dx'.$$

By paralleling the proof of Lemma 4.8.3, and replacing $\mathbf{\Gamma}$ with $\nabla \mathbf{\Gamma}$ we obtain the following result.

LEMMA 4.8.4. *If* (4.8.16) *holds then there exists a unique solution* $\nabla \mathbf{\Gamma}$ *to the problem* (4.8.17) *such that, for any* $p \in \mathbb{C}^+$,

$$\nabla \mathbf{\Gamma}(\mathbf{x}, \cdot; p) \in \overset{\sigma}{L}{}^2(\Omega), \quad (\textit{cf.}\ [120],\ \textit{Lemma}\ 23.2,$$

$$\nabla \mathbf{\Gamma}(\mathbf{x}, \mathbf{x}'; \cdot) \text{ is continuous,}$$

$$\lim_{p \to \infty} p^{1-\varepsilon} \, \nabla \mathbf{\Gamma}(\mathbf{x}, \mathbf{x}'; p) = 0, \qquad \varepsilon \in \mathbb{R}^{++}.$$

These results enable us to state existence and uniqueness of the solution to the problem (4.8.1)–(4.8.4) as follows.

THEOREM 4.8.2. *If* (4.8.14) *holds then there exists a unique solution, in* $L^2(\mathbb{R}^+; \overset{\sigma}{H}{}^1_0(\Omega))$, *to the problem* (4.8.1)–(4.8.4).

Proof. Since $\mathbf{b} \in L^2(\mathbb{R}^+; H^{-1}(\Omega))$ and $\mathbf{V}_0 \in L^2(\mathbb{R}^+; \overset{\sigma}{L}{}^2(\Omega))$, we have

$$\lim_{p \to \infty} \tilde{\mathbf{b}}(\mathbf{x}, p) = \lim_{p \to \infty} \left[\mathbf{b}_L(\mathbf{x}, p) + \mathbf{v}_0(\mathbf{x}, 0) + \int_0^\infty \exp(-ps)\mathbf{V}_0(\mathbf{x}, s) \, ds \right] = \mathbf{v}_0(\mathbf{x}, 0).$$

Then by (4.8.15) and property (iii) of $\mathbf{\Gamma}$ we obtain

$$\lim_{p \to \infty} p^{1-\varepsilon} \mathbf{v}_L(\mathbf{x}, p) = \lim_{p \to \infty} \int_\Omega p^{1-\varepsilon} \mathbf{\Gamma}(\mathbf{x}, \mathbf{x}'; p) \, \mathbf{v}(\mathbf{x}', 0) = 0. \tag{4.8.18}$$

Let $p = i\omega$, $\omega \in \mathbb{R}^+$, and $\epsilon \in (0, \frac{1}{2})$. Then (4.8.18) implies that $\mathbf{v}_L(\mathbf{x}, i\omega)$ is in $L^2(\mathbb{R}; L^2(\Omega))$ and we can view it as the Fourier transform of

$$\tilde{\mathbf{v}}(\mathbf{x}, t) = \begin{cases} 0, & t < 0, \\ \mathbf{v}(\mathbf{x}, t), & t \geq 0. \end{cases}$$

Accordingly, Plancherel's theorem allows us to write

$$
\begin{aligned}
\int_{-\infty}^{\infty}\int_{\Omega}|\mathbf{v}_L(\mathbf{x},i\omega)|^2\,dx\,d\omega &= \frac{1}{2\pi}\int_{-\infty}^{\infty}\int_{\Omega}|\tilde{\mathbf{v}}(\mathbf{x},t)|^2\,dx\,dt\\
&= \frac{1}{2\pi}\int_{0}^{\infty}\int_{\Omega}|\mathbf{v}(\mathbf{x},t)|^2\,dx\,dt. \qquad (4.8.19)
\end{aligned}
$$

By the same token we have

$$
\begin{aligned}
\int_{-\infty}^{\infty}\int_{\Omega}|\nabla\mathbf{v}_L(\mathbf{x},i\omega)|^2\,dx\,d\omega &= \frac{1}{2\pi}\int_{-\infty}^{\infty}\int_{\Omega}|\nabla\tilde{\mathbf{v}}(\mathbf{x},t)|^2\,dx\,dt\\
&= \frac{1}{2\pi}\int_{0}^{\infty}\int_{\Omega}|\nabla\mathbf{v}(\mathbf{x},t)|^2\,dx\,dt. \qquad (4.8.20)
\end{aligned}
$$

This provides the sought result $\mathbf{v} \in L^2(\mathbb{R}^+; \overset{\sigma}{H}{}_0^1(\Omega))$. □

4.9. Counterexamples to uniqueness and stability. In this section two solutions to the problem (4.5.1)–(4.5.2) are considered which, as a by-product, provide a further support of the statement in §2.5 as being the right formulation of the second law of thermodynamics. Really, the more refined version of §2.5 is characterized by the inequality being strict for any nonconstant history $\mathbf{E}^t$ and this is reflected by (3.2.11), instead of (3.4.1), being the right restriction on the relaxation function $\mathbf{G}$. Specifically, we show by two examples [64] that weakening (3.2.11) to (3.4.1) provides counterexamples to uniqueness and asymptotic stability properties of the solutions.

Consider the dynamic problem (4.5.1)–(4.5.2) and let $\mathbf{b}$ be time-harmonic, namely $\mathbf{b}(\mathbf{x},t) = \boldsymbol{\mathscr{b}}\exp(i\omega t)$. Then we look for solutions $\mathbf{u}$ in the form

$$\mathbf{u}(\mathbf{x},t) = \boldsymbol{\mathscr{u}}(\mathbf{x})\exp(i\omega t).$$

The function $\boldsymbol{\mathscr{u}}$ has to be a solution to the problem

$$\omega^2\boldsymbol{\mathscr{u}}(\mathbf{x}) + \nabla\cdot[(\mathbf{G}_0 + \mathbf{G}_F(\omega))\nabla\boldsymbol{\mathscr{u}}(\mathbf{x})] = -\boldsymbol{\mathscr{b}}(\mathbf{x}), \qquad \mathbf{x}\in\Omega, \tag{4.9.1}$$

$$\boldsymbol{\mathscr{u}}(\mathbf{x}) = 0, \qquad \mathbf{x}\in\partial\Omega. \tag{4.9.2}$$

By Theorem 4.5.2, the validity of (3.2.11) and (3.1.1) ensures existence and uniqueness of the solution to the problem (4.9.1)–(4.9.2). Now, let (3.2.11) be replaced by (3.4.1). Then there can exist at least one frequency, $\bar{\omega}$ say, such that

$$\mathbf{G}'_s(\bar{\omega}) = 0.$$

Now, to fix ideas and make the conclusion immediate, let $\Omega = \{x : x \in [0, l]\}$. Hence, as $\omega = \bar{\omega}$ the homogeneous problem becomes

$$\bar{\omega}^2 u(x) + [G_0 + G_c'(\bar{\omega})]u_{xx} = 0, \qquad x \in (0, l), \tag{4.9.3}$$

$$u(x) = 0, \qquad x = 0, l.$$

Since

$$G_0 + G_c'(\bar{\omega}) = G_\infty + \int_0^\infty G'(s)(\cos \bar{\omega} s - 1)ds \tag{4.9.4}$$

and the integral is bounded because $G' \in L^1(\mathbb{R}^+)$, we can take it that G_∞ makes $G_0 + G_c'(\omega) > 0$. Then the quantity

$$\bar{\lambda} = \frac{\rho \bar{\omega}^2}{G_0 + G_c'(\bar{\omega})}$$

is strictly positive and we can find a value $\bar{l}$ of l such that $\bar{\lambda}$ is an eigenvalue of $-d^2/dx^2$. This proves the existence of nonzero solutions to the homogeneous problem (4.83)–(4.8.4) and then nonuniqueness of solutions to the nonhomogeneous problem.

In reference to the stability properties, observe that the one-dimensional version of the problem (4.1.1)–(4.1.4) can be written as

(4.9.5)

$$u_{tt}(x,t) - G_0 u_{xx}(x,t) - \int_0^t G'(s)\, u_{xx}(x, t-s)\, ds = B(x,t) \quad \text{in } (0,l) \times \mathbb{R}^+,$$

$$u(x,t) = u_0(x,t) \quad \text{in } [0,l] \times \mathbb{R}^-, \tag{4.9.6}$$

$$u(0,t) = u(l,t) = 0, \qquad t \in \mathbb{R}^+, \tag{4.9.7}$$

where

$$B(x,t) = b(x,t) + \int_t^\infty G'(s)\, u_{0xx}(x, t-s)\, ds \in L^2(\mathbb{R}^+; L^2(0,l)).$$

Then by Theorem 4.6.1 we know that (3.1.1) and (3.2.11) imply the decay of the solution as $t \to \infty$ in the sense that $u \in L^2(\mathbb{R}^+; H_0^1((0,l)))$. If, instead, (3.4.1) holds in place of (3.2.11) then asymptotic stability may cease to hold. Letting $\bar{\omega}, \bar{l}$ be as in the previous example, denote by $\bar{w} \in H^2((0,l))$ any nonzero solution to (4.9.3)–(4.9.4). The choice $b = 0$ makes $u(x,t) =$

$\bar{w}\exp(i\bar{\omega}t)$, $x \in (0,l)$, $t \in \mathbb{R}^{++}$, with $u_0(x,t) = \bar{w}\exp(i\bar{\omega}t)$ in $[0,l] \times \mathbb{R}^-$, a solution to (4.9.5)–(4.9.7). Of course such $u(x,t)$ does not decay as $t \to \infty$.

Incidentally, if also $G_0 + G_c'(\bar{\omega}) = 0$ then by (4.9.3) even the existence of the solution ceases to hold.

Analogous results hold for fluids. Assume that an incompressible viscoelastic fluid occupies the region $\{(x,y,z) : x \in [0,l],\ y,z \in \mathbb{R}\}$ and the velocity depends on x only and is directed along an orthogonal direction, y say. Then the corresponding one-dimensional problem for the velocity v has the form

(4.9.8)
$$v_t(x,t) = \int_0^\infty \mu(s) v_{xx}(x,t-s)\, ds + b(x,t), \qquad x \in (0,l), \qquad t \in \mathbb{R}^{++},$$

(4.9.9)
$$v(0,t) = v(l,t) = 0, \qquad t \in \mathbb{R}^+,$$

(4.9.10)
$$v(x,\tau) = v_0(x,\tau), \qquad x \in (0,l), \qquad \tau \in \mathbb{R}^-.$$

We show that replacing (3.7.5) with the weaker condition $\mu_c(\omega) \geq 0$, $\omega \in \mathbb{R}^{++}$, implies that asymptotic stability may cease to hold. For technical reasons we let μ be monotonically decreasing on $\mathbb{R}^+$. Then we prove the following theorem.

THEOREM 4.9.1. *Let $\mu_c(\bar{\omega}) = 0$ for some $\bar{\omega} \in \mathbb{R}^{++}$. Then there exists a critical length $\bar{l}$ such that for a given initial history $v_0(x,\tau) = \sin(\pi x/l)$ $[c_1 \cos\bar{\omega}\tau + c_2 \sin\bar{\omega}\tau]$, $c_1, c_2 \in \mathbb{R}$, and vanishing body force, the problem (4.9.8)–(4.9.10) has a nontrivial solution.*

Proof. Let $v(x,t) = V(x)\exp(i\bar{\omega}t)$. By (4.9.8) and (4.9.9) we have

(4.9.11)
$$\bar{\omega}V + \mu_s(\bar{\omega})V_{xx} = 0, \qquad V(0) = V(l) = 0.$$

Because

(4.9.12)
$$\frac{1}{\bar{\omega}}\mu_s(\bar{\omega}) = \frac{1}{\bar{\omega}^2}\left[\mu_0 + \int_0^\infty \mu'(s)\cos\bar{\omega}s\, ds\right] = \frac{1}{\bar{\omega}^2}\int_0^\infty \mu'(s)[\cos\bar{\omega}s - 1]ds > 0$$

we can define the length

$$\bar{l} = \pi\sqrt{\frac{\mu_s(\bar{\omega})}{\bar{\omega}}}.$$

Then the function $V(x) = c\sin(\pi x/\bar{l})$, $c \in \mathbb{R}$, is a solution to the problem (4.9.11). Now it is easy to see that, if $b = 0$,

$$v(x,t) = \sin\left(\frac{\pi x}{\bar{l}}\right)[c_1 \cos\bar{\omega}t + c_2 \sin\bar{\omega}t]$$

is the unique, nontrivial solution to the problem (4.9.8)–(4.9.10). □

Theorem 4.9.1 means that, with $b = 0$, asymptotic stability ceases to hold while the initial history $v_0(x, \tau)$ is bounded on $[0, \bar{l}] \times \mathbb{R}^-$. Incidentally, this proves also that uniqueness does not hold as $b \neq 0$.

The hypothesis of Theorem 4.9.1 is exemplified by considering the relaxation function

$$\mu(s) = \left(s^2 - \frac{\nu - 3}{\upsilon} s + \frac{\nu^2 - 8\nu + 4}{8\upsilon^2} \right) \exp(-\upsilon s)$$

where $\nu \in (2 - \sqrt{2},\, 2 + \sqrt{2}),\ \upsilon > 0$. It is easily seen that $\mu(s) > 0,\ \mu'(s) < 0,\ s \in [0, \infty)$. Moreover,

$$\mu_c(\omega) = \frac{\nu^2}{8\upsilon(\upsilon^2 + \omega^2)^3} \left(\omega^2 - \frac{8 - \nu}{\nu} \upsilon^2 \right)^2 .$$

This means that $\mu_c \geq 0$ and $\mu_c = 0$ at $\omega = \sqrt{(8 - \nu)/\nu}\, \upsilon$.

Chapter 5

Variational Formulations and Minimum Properties

5.1. Preliminaries on variational formulations. As in other fields of continuum mechanics, there are many reasons which motivate the recourse to variational formulations. Essentially, a single functional incorporates all of the intrinsic features of the problem at hand: differential equations, boundary and jump conditions. Also, the knowledge of a functional allows a systematic connection between symmetries and conservation laws (cf. [11] and [100], Chap. 11). Moreover, variational formulations constitute the starting point of many approximation methods, both analytical and numerical.

Among the procedures for the derivation of variational formulations, the most traditional one is patterned after Hamilton's principle. Very often, however, the derivation of a Lagrangian as for particle-like systems, viz. $\mathcal{L} = T - V$, is out of the question merely because usually the concept of potential energy is not operatively clear. Things are even more difficult when materials with memory are concerned. That is why we need a systematic method for the derivation of variational formulations. The systematic method par excellence is that pertaining to the theory of the inverse problem of the calculus of variations. While its application to viscoelasticity is developed in the next section, here we state the main results of the theory which are relevant to our purpose.

Let X be a Banach space over $\mathbb{R}$ and X^* denote the conjugate (dual) space of X. Letting $v \in X^*$ and $z \in X$, $\langle v, z\rangle$ represents a nondegenerate bilinear form whereby if $\langle v, z\rangle = 0$ for every $v \in X^*$ $(z \in X)$ then z (v) is the null element of X (X^*). For each subset U of X we let $\mathcal{N}(U; X^*)$ denote the vector space of all operators from U into X^*. Suppose that $U \subset X$ is an open set and let $N \in \mathcal{N}(U; X^*)$. Let $u + \nu h \in U, \nu \in [0, \bar{\nu}]$, for some $\bar{\nu} \in \mathbb{R}^{++}$. If

N is Gâteaux differentiable we denote by

$$dN(u \mid h) = \left.\frac{dN(u+\nu h)}{d\nu}\right|_{\nu=0}$$

the (Gâteaux) differential of N at $u \in U$ in the direction h. If there exists $N'(u)$ such that

$$dN(u \mid h) = \langle N'(u), h\rangle,$$

we regard $N'(u)$ as the derivative of N with respect to the bilinear form. Similarly, in the case of functionals $f : X \to \mathbb{R}$ we write $df(u \mid h) = \langle f'(u), h\rangle$. If there exists a functional f such that $N = f'$, then we say that N is a potential operator. These preliminaries are enough to understand the content of the following theorem whose proof is given by Vainberg[27] [123] (cf. also [1]) and then is omitted.

THEOREM 5.1.1. *Suppose that*

(1) *N is an operator from X into X^*;*

(2) *N has a linear Gâteaux differential $dN(u \mid h)$ at every point of the open convex set $D \subset X$;*

(3) *the bilinear form $\langle dN(u \mid h), k\rangle$ on $h, k \in X$ is continuous in u at every point of D.*

Then a necessary and sufficient condition for N to be a potential operator in D is that

$$\langle dN(u \mid h), k\rangle = \langle dN(u \mid k), h\rangle \tag{5.1.1}$$

for every $h, k \in X$ and every $u \in D$. Moreover, if (5.1.1) *holds then $N = f'$ with*

$$f(u) = f(u_0) + \int_0^1 \langle N(u_0 + \nu(u - u_0)), u - u_0\rangle \, d\nu, \tag{5.1.2}$$

u_0 being any point of D.

Most often variational formulations are established by letting X be the Hilbert space $L^2(\mathcal{D})$, $\mathcal{D}$ being a suitable domain, and $\langle\cdot,\cdot\rangle$ be the scalar product in $L^2(\mathcal{D})$. In connection with viscoelasticity, and more generally in the case of constitutive functionals expressed by convolutions, this choice is usually unsuccessful. Since the bilinear form is required to be nondegenerate, an alternative choice for $\langle\cdot,\cdot\rangle$ in the case of space-time-dependent functions is the L^2 scalar product with respect to space variables and the convolution with respect to time.

---

[27] Although stated in a less clear form, this theorem was first established by Volterra [127].

Let $X = L^2(\Omega \times I)$, Ω being a region of the three-dimensional Euclidean space and $I = [0,t]$ for some $t \in \mathbb{R}^{++}$. Then for any pair of functions $v, z \in L^2(\Omega \times I)$ we set

$$\langle v, z \rangle = \int_\Omega \int_0^t v(\tau)\, z(t-\tau) d\tau\, dx, \tag{5.1.3}$$

where the dependence of v, z on $\mathbf{x} \in \Omega$ is understood and not written. Not to be pedantic, we omit indicating explicitly that the bilinear form $\langle \cdot, \cdot \rangle$ depends on (the parameter) t. For ease in writing we denote by $*$ the convolution operator, namely

$$v * z = \int_0^t v(\tau)\, z(t-\tau)\, d\tau.$$

If v, z have values in finite-dimensional inner-product vector spaces, then the product $v\,z$ in (5.1.3) is meant as the corresponding inner product. About the convolution (5.1.3), we recall Titchmarsh's theorem whereby

$$v * z \equiv 0 \Rightarrow \text{either} \quad v = 0 \quad \text{or} \quad z = 0 \quad \text{for any} \quad t \geq 0.$$

This means that the convolution constitutes a nondegenerate bilinear form.

On the basis of (5.1.1)-(5.1.3), in the next sections we set up some variational formulations for the dynamic problem of viscoelasticity. Here we remark that once the bilinear form (5.1.3) is adopted, any system of equations admits a variational formulation only if the spatial derivatives satisfy the potentialness conditions in L^2. These conditions are by now well known; the interested reader is referred to [119], [1], [2] for a detailed account.

5.2. Variational formulation via the theory of the inverse problem. The dynamic problem in viscoelasticity consists in determining the solution $\mathbf{u}(\mathbf{x}, t) \in C^{2,2}(\Omega \times \mathbb{R}^{++}, V)$ when $\mathbf{u}(\mathbf{x}, t)$ is known in $\Omega \times \mathbb{R}^-$ along with suitable boundary values.[28] To be specific, we write the stress tensor $\mathbf{T}$ as

$$\mathbf{T}(t) = \mathbf{G}_0\, \mathbf{E}(t) + \int_0^t \mathbf{G}'(s)\, \mathbf{E}(t-s) ds + \mathbf{T}_0(t), \qquad t \in \mathbb{R}^+,$$

with

$$\mathbf{T}_0(t) = \int_t^\infty \mathbf{G}'(s)\, \mathbf{E}(t-s)\, ds.$$

Letting L be the linear operator defined as

$$L\mathbf{u} := \rho\ddot{\mathbf{u}} - \nabla \cdot (\mathbf{G}_0 \nabla \mathbf{u} + \mathbf{G}' * \nabla \mathbf{u}) - \rho \mathbf{f}$$

[28] Henceforth a symbol $C^{a,b}$ means of class C^a with respect to (the first variable) $\mathbf{x}$ and of class C^b with respect to t.

where

$$\rho\mathbf{f} = \rho\mathbf{b} + \nabla \cdot \mathbf{T}_0,$$

we can write the equation of motion as

$$L\mathbf{u} = 0. \tag{5.2.1}$$

To see whether (5.2.1) admits a variational formulation, we consider the convolution bilinear form (5.1.3). The potentialness condition (5.1.1) for $N = L$ holds if and only if

$$L\mathbf{w} * \mathbf{u} = L\mathbf{u} * \mathbf{w} \tag{5.2.2}$$

holds. Suitable integrations by parts show that (5.2.2) holds (to within initial-boundary terms) by virtue of the commutativity of the convolution and the symmetry of $\mathbf{G}_0$ and $\mathbf{G}'(s), s \in \mathbb{R}^+$. Then the application of (5.1.2) leads to the corresponding functional in the form

$$\Lambda(\mathbf{u}) = \int_\Omega [\tfrac{1}{2}\rho\dot{\mathbf{u}} * \dot{\mathbf{u}} + \tfrac{1}{2}\nabla\mathbf{u} * (\mathbf{G}_0\nabla\mathbf{u} + \mathbf{G}' * \nabla\mathbf{u}) - \rho\mathbf{u} * \mathbf{f}]\, dx.$$

A variational formulation must incorporate the initial-boundary conditions. Let the boundary $\partial\Omega$ of Ω consist of two disjoint subsets $\partial\Omega_u, \partial\Omega_T$ such that

$$\partial\Omega_u \cup \partial\Omega_T = \partial\Omega, \qquad \partial\Omega_u \cap \partial\Omega_T = \emptyset.$$

Consider the initial-boundary value problem (P)

$$L\mathbf{u} = 0 \quad \text{on } \Omega \times \mathbb{R}^{++},$$

$$\mathbf{u} = \hat{\mathbf{u}} \quad \text{on } \partial\Omega_u \times \mathbb{R}^+, \qquad \mathbf{T}\,\mathbf{n} = \hat{\mathbf{t}} \quad \text{on } \partial\Omega_T \times \mathbb{R}^+, \tag{5.2.3}$$

$$\mathbf{u}(0) = \mathbf{u}_0, \qquad \dot{\mathbf{u}}(0) = \dot{\mathbf{u}}_0 \quad \text{on } \bar{\Omega}, \tag{5.2.4}$$

and let $\mathbf{T}_0$ be given on $\bar{\Omega} \times \mathbb{R}^+$. The functional

$$\begin{aligned}\bar{\Lambda}(\mathbf{u}) = &\int_\Omega [\tfrac{1}{2}\rho\dot{\mathbf{u}} * \dot{\mathbf{u}} + \tfrac{1}{2}\nabla\mathbf{u} * (\mathbf{G}_0\, \nabla\mathbf{u} + \mathbf{G}' * \nabla\mathbf{u}) - \rho\mathbf{u} * \mathbf{f} - \rho\mathbf{u} \cdot \dot{\mathbf{u}}_0]\, dx \\ &- \int_{\partial\Omega_T} \mathbf{u} * (\hat{\mathbf{t}} - \mathbf{T}_0\mathbf{n})\, da\end{aligned}$$

provides a variational formulation for the problem P as is shown through the following theorem.

THEOREM 5.2.1. *Let K be the set of functions in $C^{2,2}(\bar{\Omega} \times \mathbb{R}^+, V)$ which meet the conditions* $(5.2.3)_1$ *and* $(5.2.4)_1$. *Then the functional $\bar{\Lambda}$ has a stationary point $\mathbf{u} \in K$ if and only if $\mathbf{u}$ is a solution to the mixed problem* P.

Proof. As any functional of convolution type, $\bar{\Lambda}$ is parameterized by the time $t \in \mathbb{R}^+$ as the upper limit of integration. Let $\mathbf{w} \in K$ vanish on $\partial\Omega_u \times \mathbb{R}^+$ and on $\bar{\Omega}$ at $t = 0$. Then $\mathbf{u} + \nu\mathbf{w} \in K$ for any real ν if $\mathbf{u} \in K$. Owing to the symmetry of $\mathbf{G}_0$ and $\mathbf{G}'$ and the commutativity property of the convolution, we have

$$\begin{aligned} d\bar{\Lambda}(\mathbf{u} \mid \mathbf{w}) = & \int_\Omega \int_0^t \rho\dot{\mathbf{w}}(\tau) \cdot \dot{\mathbf{u}}(t-\tau)\, d\tau\, dx \\ & + \int_\Omega \int_0^t \nabla\mathbf{w}(\tau) \cdot \mathbf{G}_0 \nabla\mathbf{u}(t-\tau)\, d\tau\, dx \\ & + \int_\Omega \int_0^t \int_0^{t-\tau} \nabla\mathbf{w}(\tau) \cdot \mathbf{G}'(t-\tau-\xi)\nabla\mathbf{u}(\xi)\, d\xi\, d\tau\, dx \\ & - \int_\Omega \int_0^t \rho\mathbf{w}(\tau) \cdot \mathbf{f}(t-\tau) d\tau\, dx \;-\; \int_\Omega \rho\mathbf{w}(t) \cdot \dot{\mathbf{u}}_0\, dx \\ & - \int_{\partial\Omega_T} \int_0^t \mathbf{w}(\tau) \cdot [\hat{\mathbf{t}}(t-\tau) - \mathbf{T}_0(t-\tau)\mathbf{n}] da. \end{aligned}$$

Integration by parts with respect to τ in the first integral and with respect to space variables in the second and third ones gives

$$\begin{aligned} d\bar{\Lambda}(\mathbf{u} \mid \mathbf{w}) = & \int_\Omega \int_0^t \mathbf{w}(\tau) \cdot [\rho\ddot{\mathbf{u}} - \nabla \cdot (\mathbf{G}_0 \nabla\mathbf{u} + \mathbf{G}' * \nabla\mathbf{u}) - \rho\mathbf{f}](t-\tau) d\tau dx \\ & + \int_\Omega \rho\mathbf{w}(t) \cdot [\dot{\mathbf{u}}(0) - \dot{\mathbf{u}}_0]\, dx \\ & + \int_{\partial\Omega_T} \int_0^t \mathbf{w}(\tau) \cdot [-\hat{\mathbf{t}} + (\mathbf{T}_0 + \mathbf{G}_0\, \nabla\mathbf{u} + \mathbf{G}' * \nabla\mathbf{u})\, \mathbf{n}](t-\tau)\, d\tau\, da. \end{aligned}$$

The arbitrariness of $\mathbf{w}$ on $\Omega \times \mathbb{R}^{++}$ and $\partial\Omega_T \times \mathbb{R}^{++}$ proves the "only if" part.[29] The proof of the "if" part is obvious. □

5.3. Further variational formulations. The variational formulation in § 5.2 is the simplest and most direct version for the mixed problem P; it is the result of the procedure pertaining to the inverse problem and involves one unknown (vector) function only, namely the displacement field $\mathbf{u}$. Variational formulations with a minimal number of unknown functions seem to be the

[29] Henceforth we apply some analogues of the fundamental lemma of the calculus of variations for the convolution in $\Omega, \partial\Omega_u, \partial\Omega_T$ (cf. [76], [92]).

most appropriate ones for applications, especially in connection with numerical approximations of the solution. Nonetheless, it may be of interest, at least from a theoretical viewpoint, to set up variational formulations which incorporate the equation of motion, the initial-boundary conditions, the constitutive equations, and possibly kinematic relations. The results of this section give a unified version of previous theorems, due to various authors, and provide generalizations to the dynamic context of some theorems by Gurtin [74] about quasi-static viscoelasticity.

The mixed problem of viscoelasticity can be written in the following form, denoted as problem P_ξ. While the initial-boundary conditions are those of the problem P, the system of equations is written as

$$-\mathbf{E} + \mathrm{sym}\nabla\mathbf{u} = 0,$$

$$-\tilde{\mathbf{T}} + \mathbf{G}_0\,\mathbf{E} + \mathbf{G}' * \mathbf{E} = 0,$$

$$\rho\ddot{\mathbf{u}} - \nabla\cdot\tilde{\mathbf{T}} - \rho\mathbf{f} = 0$$

where $\tilde{\mathbf{T}} = \mathbf{T} - \mathbf{T}_0$. Let ξ be the triplet of unknown functions $(\tilde{\mathbf{T}}, \mathbf{E}, \mathbf{u})$ and let L_ξ be the linear differential operator associated with the system of equations. Then the system of equations admits a variational formulation if and only if

$$\langle L_\xi\xi_1, \xi_2\rangle = \langle L_\xi\xi_2, \xi_1\rangle$$

for any pair of admissible triplets ξ_1, ξ_2. Such turns out to be the case and hence the corresponding functional is evaluated through (5.1.2). Once the appropriate initial-boundary terms are determined, we obtain the following variational formulation.[30]

THEOREM 5.3.1. *Let H be the set of all triplets $\xi = (\tilde{\mathbf{T}}, \mathbf{E}, \mathbf{u})$ with $\tilde{\mathbf{T}}, \mathbf{E} : \Omega \times \mathbb{R}^+ \to \mathrm{Sym}$, of class $C^{1,0}$ and $C^{0,0}$, respectively, and $\mathbf{u} : \Omega \times \mathbb{R}^+ \to V$ of class $C^{1,2}$. Moreover let the elements ξ of H satisfy the initial-boundary conditions* $(5.2.3)_1$ *and* $(5.2.4)_1$. *Then the functional*

$$\Xi(\xi) = \int_\Omega [-\mathbf{E} * \tilde{\mathbf{T}} + \nabla\mathbf{u} * \tilde{\mathbf{T}} + \tfrac{1}{2}\mathbf{E} * (\mathbf{G}_0\,\mathbf{E} + \mathbf{G}' * \mathbf{E}) + \tfrac{1}{2}\rho\dot{\mathbf{u}} * \dot{\mathbf{u}} - \rho\mathbf{f} * \mathbf{u}]\,dx$$
$$- \int_\Omega \rho\mathbf{u}\cdot\dot{\mathbf{u}}_0\,dx - \int_{\partial\Omega_T} \mathbf{u} * (\hat{\mathbf{t}} - \mathbf{T}_0\mathbf{n})\,da$$

has a stationary point ξ in H if and only if ξ is a solution to the mixed problem P_ξ.

[30] A similar variational formulation is established by Reddy [109] by having recourse to Stieltjes convolution.

Proof. Let $\eta = (\boldsymbol{\tau}, \boldsymbol{\epsilon}, \mathbf{w}) \in H$ with $\mathbf{w} = 0$ on $\partial\Omega_u \times \mathbb{R}^+$ and on $\bar{\Omega}$ at $t = 0$. Then $\xi + \nu\eta \in H$ for any real ν. By the commutativity of the convolution and the symmetry of $\mathbf{G}_0$ and $\mathbf{G}'$ we have

$$d\Xi(\xi \mid \eta) = \int_\Omega [-\boldsymbol{\epsilon} * \tilde{\mathbf{T}} - \mathbf{E} * \boldsymbol{\tau} + \nabla\mathbf{w} * \tilde{\mathbf{T}} + \nabla\mathbf{u} * \boldsymbol{\tau} + \boldsymbol{\epsilon} * (\mathbf{G}_0\, \mathbf{E} + \mathbf{G}' * \mathbf{E})$$
$$+ \rho\dot{\mathbf{w}} * \dot{\mathbf{u}} - \rho\mathbf{w} * \mathbf{f}]\, dx$$
$$- \int_\Omega \rho\mathbf{w} \cdot \dot{\mathbf{u}}_0\, dx - \int_{\partial\Omega_T} \mathbf{w} * (\hat{\mathbf{t}} - \mathbf{T}_0\mathbf{n})\, da.$$

Integration by parts of $\nabla\mathbf{w} * \tilde{\mathbf{T}}$ with respect to space variables and of $\rho\dot{\mathbf{w}} * \dot{\mathbf{u}}$ with respect to time gives

$$d\Xi(\xi \mid \eta) = \int_\Omega [\boldsymbol{\tau} * (\mathrm{sym}\nabla\mathbf{u} - \mathbf{E}) + \boldsymbol{\epsilon} * (-\tilde{\mathbf{T}} + \mathbf{G}_0\, \mathbf{E} + \mathbf{G}' * \mathbf{E})$$
$$+ \mathbf{w} * (\rho\ddot{\mathbf{u}} - \nabla \cdot \tilde{\mathbf{T}} - \rho\mathbf{f})]\, dx$$
$$+ \int_\Omega \rho\mathbf{w} \cdot [\dot{\mathbf{u}}(0) - \dot{\mathbf{u}}_0]\, dx + \int_{\partial\Omega_T} \mathbf{w} * [(\tilde{\mathbf{T}} + \mathbf{T}_0)\mathbf{n} - \hat{\mathbf{t}}]\, da.$$

The arbitrariness of $\boldsymbol{\tau}$, $\boldsymbol{\epsilon}$, and $\mathbf{w}$ on $\Omega \times \mathbb{R}^{++}$ and of $\mathbf{w}$ on $\partial\Omega_T \times \mathbb{R}^{++}$ proves the "only if" part. The proof of the "if" part is obvious. □

Theorems 5.2.1 and 5.3.1 are related to the functions $\mathbf{u}$ which satisfy the initial-boundary conditions $(5.2.3)_1$ and $(5.2.4)_1$. It is possible to incorporate these conditions in a variational formulation thus searching for the stationary point without any significant restriction on the set of admissible functions.

This is accomplished through the next theorem which may be viewed as the counterpart of that established by Herrera and Bielak ([80], Theorem 8.2) in elastodynamics. The proof involves the concept of nonsingular operator in the following way.

DEFINITION 5.3.1. An operator $M : V \to \mathbb{R}$ is called *nonsingular* if for every vector $\mathbf{v} \in V$ we have

$$M\mathbf{v} = 0 \Rightarrow \mathbf{v} = 0.$$

Consider the vector field $\mathbf{a} = \mathbf{n}(\mathbf{G}_0\, \nabla\mathbf{w} + \mathbf{G}' * \nabla\mathbf{w})$ where the unit normal $\mathbf{n}$ and $\nabla\mathbf{w}$ are continuous on $\partial\Omega_u \times \mathbb{R}^+$. Since $\mathbf{G}_0$ is a positive-definite tensor in Sym, for any time $t \in \mathbb{R}^+$ it is always possible to choose histories of $\nabla\mathbf{w}$ in $[0, t]$ such that $\nabla\mathbf{w}(t)$ is arbitrary while $\mathbf{G}' * \nabla\mathbf{w}$ is as small as we please, relative to the norm of Sym. Let $M\mathbf{v} = \mathbf{v} \otimes \mathbf{n}\; \mathbf{G}_0\; \mathbf{H}$, $\mathbf{H} \in$ Sym. Owing to the arbitrariness of $\mathbf{H}$ and the positive definiteness of $\mathbf{G}_0$, $M\mathbf{v} = 0$ always implies that sym $\mathbf{v} \otimes \mathbf{n} = 0$, which in turn implies that $\mathbf{v} = 0$. In conclusion, the inner product with $\mathbf{a}(\mathbf{w})$, for any admissible $\mathbf{w}$, is a nonsingular operator.

THEOREM 5.3.2. *Let* $\mathbf{u} \in C^{2,2}(\Omega \times \mathbb{R}^+, V)$. *The functional*

$$\Lambda_0(\mathbf{u}) = \int_\Omega [\tfrac{1}{2}\rho\dot{\mathbf{u}} * \dot{\mathbf{u}} + \tfrac{1}{2}\nabla\mathbf{u} * (\mathbf{G}_0\,\nabla\mathbf{u} + \mathbf{G}' * \nabla\mathbf{u}) - \rho\mathbf{u} * \mathbf{f}]\,dx$$
$$+ \int_\Omega \rho[(\mathbf{u}(0) - \mathbf{u}_0)\cdot\dot{\mathbf{u}} - \dot{\mathbf{u}}_0\cdot\mathbf{u}]\,dx$$
$$+ \int_{\partial\Omega_u} (\hat{\mathbf{u}} - \mathbf{u}) * (\mathbf{G}_0\,\nabla\mathbf{u} + \mathbf{G}' * \nabla\mathbf{u})\mathbf{n}\,da$$
$$- \int_{\partial\Omega_T} (\hat{\mathbf{t}} - \mathbf{T}_0\mathbf{n}) * \mathbf{u}\,da$$

is stationary at $\mathbf{u}$ *if and only if* $\mathbf{u}$ *is a solution of the mixed problem* P.

Proof. Let $\mathbf{w} \in C^{2,2}(\Omega\times\mathbb{R}^+, V)$. By proceeding along the lines of Theorem 5.2.1, we obtain

$$d\Lambda_0(\mathbf{u} \mid \mathbf{w}) = \int_\Omega \mathbf{w} * [\rho\ddot{\mathbf{u}} - \nabla\cdot(\mathbf{G}_0\,\nabla\mathbf{u} + \mathbf{G}' * \nabla\mathbf{u}) - \rho\mathbf{f}]\,dx$$
$$+ \int_\Omega [\rho\mathbf{w}\cdot(\dot{\mathbf{u}}(0) - \dot{\mathbf{u}}_0) + \rho(\mathbf{u}(0) - \mathbf{u}_0)\cdot\dot{\mathbf{w}}]\,dx$$
$$+ \int_{\partial\Omega_u} \mathbf{n}(\mathbf{G}_0\,\nabla\mathbf{w} + \mathbf{G}' * \nabla\mathbf{w}) * (\hat{\mathbf{u}} - \mathbf{u})\,da$$
$$+ \int_{\partial\Omega_T} \mathbf{w} * [(\mathbf{T}_0 + \mathbf{G}_0\,\nabla\mathbf{u} + \mathbf{G}' * \nabla\mathbf{u})\mathbf{n} - \hat{\mathbf{t}}]\,da.$$

First consider $\mathbf{w} = 0$ on $\partial\Omega_T$ and $\nabla\mathbf{w} = 0$ on $\partial\Omega_u$, identically in the time interval under consideration, but otherwise arbitrary. Then $d\Lambda_0(\mathbf{u} \mid \mathbf{w}) = 0$ only if the equation of motion and the initial conditions hold. Now we let $\mathbf{w}$ be arbitrary on $\partial\Omega_T$ but $\nabla\mathbf{w} = 0$ on $\partial\Omega_u$. Hence $d\Lambda_0$ vanishes only if the boundary condition for $\mathbf{T}$ holds. Finally, let $\nabla\mathbf{w}$ be arbitrary on $\partial\Omega_u$. Since $(\mathbf{G}_0\,\nabla\mathbf{w} + \mathbf{G}' * \nabla\mathbf{w})$ is nonsingular, it follows that $\mathbf{u} = \hat{\mathbf{u}}$ must hold on $\partial\Omega_u$, which completes the "only if" part. That any $\mathbf{u}$ satisfying P makes Λ_0 be stationary is evident. □

Of course an analogous, direct generalization of Theorem 5.3.1 holds. Here, though, we prefer to give a variational formulation that does not involve time derivatives of the unknown function $\mathbf{u}$. This formulation is based on a procedure developed by Gurtin in connection with linear initial-value problems [75] and linear elastodynamics [76]. In the literature, formulations based on such a procedure, where the given differential equation is transformed to a convolution form, are usually said to be of "Gurtin-type."

Observe that, as shown by Ignaczak [83], the equation of motion can be written in terms of convolutions and without time derivatives. In our notation, it is

$$\rho\mathbf{u} = g * \nabla\cdot\tilde{\mathbf{T}} + \boldsymbol{\mathcal{F}}, \qquad t \in \mathbb{R}^+,$$

where

$$g(t) = t,$$

$$\boldsymbol{\mathcal{F}}(\mathbf{x}, t) = [g * (\rho\mathbf{b} + \nabla \cdot \mathbf{T}_0)](\mathbf{x}, t) + \rho[t\dot{\mathbf{u}}(\mathbf{x}, 0) + \mathbf{u}(\mathbf{x}, 0)].$$

Accordingly, the initial conditions are embodied in the equation of motion and then the mixed problem of viscoelasticity can be given the following form (problem $\mathrm{P_c}$). The triplet $\xi = (\tilde{\mathbf{T}}, \mathbf{E}, \mathbf{u})$ of unknown functions meets the system of equations

$$-\mathbf{E} + \mathrm{sym}\nabla\mathbf{u} = 0,$$

$$-\tilde{\mathbf{T}} + \mathbf{G}_0\,\mathbf{E} + \mathbf{G}' * \mathbf{E} = 0,$$

$$\rho\mathbf{u} - g * \nabla \cdot \tilde{\mathbf{T}} - \boldsymbol{\mathcal{F}} = 0$$

on $\Omega \times \mathbb{R}^+$ and the boundary conditions

$$\mathbf{u} = \hat{\mathbf{u}} \quad \text{on } \partial\Omega_u \times \mathbb{R}^+, \qquad \mathbf{Tn} = \hat{\mathbf{t}} \quad \text{on } \partial\Omega_T \times \mathbb{R}^+$$

where $\mathbf{T} = \tilde{\mathbf{T}} + \mathbf{T}_0$. Of course $\boldsymbol{\mathcal{F}}$ and $\mathbf{T}_0$ are regarded as known functions on $\Omega \times \mathbb{R}^+$.

Owing to the appearance of $g*$ in the third equation, the problem $\mathrm{P_c}$ as such does not meet the potentialness condition, relative to the bilinear form (5.1.3). Upon applying the convolution with g to the first and second equation, the potentialness condition is satisfied. Accordingly we determine the desired functional and add the appropriate boundary terms. The variational formulation is made precise through the following theorem.

THEOREM 5.3.3. *Let H_c be the set of all triplets $\xi = (\tilde{\mathbf{T}}, \mathbf{E}, \mathbf{u})$ with $\tilde{\mathbf{T}}, \mathbf{E} : \Omega \times \mathbb{R}^+ \to \mathrm{Sym}$, of class $C^{1,0}$ and $C^{0,0}$, respectively, and $\mathbf{u} : \Omega \times \mathbb{R}^+ \to V$ of class $C^{1,2}$. The functional*

$$\begin{aligned}\Xi_c(\xi) = \int_\Omega [\tfrac{1}{2} g * \mathbf{E} * (\mathbf{G}_0\mathbf{E} + \mathbf{G}' * \mathbf{E}) + \tfrac{1}{2}\rho\mathbf{u} * \mathbf{u} \\ - g * \tilde{\mathbf{T}} * \mathbf{E} - (g * \nabla \cdot \tilde{\mathbf{T}} + \boldsymbol{\mathcal{F}}) * \mathbf{u}]\, dx \\ + \int_{\partial\Omega_u} g * \tilde{\mathbf{T}}\mathbf{n} * \hat{\mathbf{u}}\, da + \int_{\partial\Omega_T} g * [(\tilde{\mathbf{T}} + \mathbf{T}_0)\mathbf{n} - \hat{\mathbf{t}}] * \mathbf{u}\, da\end{aligned}$$

is stationary at a point ξ in H_c if and only if ξ is a solution to the mixed problem $\mathrm{P_c}$.

Proof. Letting $l(t) = 1,\ t \in \mathbb{R}^+$, we have

$$l * \mathbf{E} * \mathbf{G} * \mathbf{E} = g * \mathbf{E} * (\mathbf{G}_0\mathbf{E} + \mathbf{G}' * \mathbf{E}).$$

This allows us to identify the functional Ξ_c with the functional in [92], Theorem 4.1, and to apply the same proof. □

5.4. Variational formulations for problems of creep type. Similar variational formulations hold when the stress is regarded as an independent variable in the sense that the constitutive relation is expressed by the creep law

$$\mathbf{E}(t) = \mathbf{J}_0\mathbf{T}(t) + \int_0^t \mathbf{J}'(s)\mathbf{T}(t-s)\, ds \; + \mathbf{E}_0(t)$$

where

$$\mathbf{E}_0(t) = \int_t^\infty \mathbf{J}'(s)\mathbf{T}(t-s)\, ds.$$

It is quite natural to set up the counterparts of Theorems 5.3.1 and 5.3.3. Instead, it will be much less obvious to arrive at the counterparts of Theorems 5.2.1 and 5.3.2 due to the difficulty of writing all equations in terms of the stress tensor $\mathbf{T}$ only.

The strict analogue of Theorem 5.3.1 would be one where the triplet $(\mathbf{T}, \tilde{\mathbf{E}}, \mathbf{u})$, with $\tilde{\mathbf{E}} = \mathbf{E} - \mathbf{E}_0$, is the set of independent functions. In such a case, however, we cannot find a variational formulation. Regard $\zeta = (\mathbf{T}, \mathbf{u})$ as the pair of unknown functions and express the corresponding mixed problem P_ζ as follows. The pertinent equations are

$$\mathrm{sym}\nabla\mathbf{u} - \mathbf{E}_0 - \mathbf{J}_0\mathbf{T} - \mathbf{J}' * \mathbf{T} = 0,$$

$$\rho\ddot{\mathbf{u}} - \nabla\cdot\mathbf{T} - \rho\mathbf{b} = 0$$

and the associated initial-boundary conditions are those of the problem P. Relative to the convolution bilinear form (5.1.3), this system of equations meets the potentialness condition (5.1.1). Then we determine the corresponding functional to be supplemented with the appropriate initial-boundary terms. As a result we have the following formulations which incorporate the initial-boundary conditions.

THEOREM 5.4.1. *Let $\tilde{H}$ be the set of all pairs $\zeta = (\mathbf{T}, \mathbf{u})$ with $\mathbf{T} : \Omega \times \mathbb{R}^+ \to$ Sym of class $C^{1,0}$ and $\mathbf{u} : \Omega \times \mathbb{R}^+ \to V$ of class $C^{1,2}$. Then the functional*

$$\begin{aligned}\Upsilon(\zeta) = &\int_\Omega [(\mathrm{sym}\nabla\mathbf{u} - \mathbf{E}_0) * \mathbf{T} - \tfrac{1}{2}\mathbf{T} * (\mathbf{J}_0\mathbf{T} + \mathbf{J}' * \mathbf{T}) + \tfrac{1}{2}\rho\dot{\mathbf{u}} * \dot{\mathbf{u}} - \rho\mathbf{b} * \mathbf{u}]\, dx \\ &+ \int_\Omega \rho[(\mathbf{u}(0) - \mathbf{u}_0)\cdot\dot{\mathbf{u}} - \dot{\mathbf{u}}_0\cdot\mathbf{u}]\, dx + \int_{\partial\Omega_u} (\hat{\mathbf{u}} - \mathbf{u}) * \mathbf{T}\mathbf{n}\, da \\ &- \int_{\partial\Omega_T} \hat{\mathbf{t}} * \mathbf{u}\, da\end{aligned}$$

is stationary at $\zeta \in \tilde{H}$ if and only if ζ is a solution to the mixed problem P_ζ.

Proof. Let $\omega = (\boldsymbol{\tau}, \mathbf{w}) \in \tilde{H}$. By means of suitable integrations by parts we obtain

$$d\Upsilon(\zeta \mid \omega) = \int_{\Omega} \mathbf{w} * [\rho\ddot{\mathbf{u}} - \nabla \cdot \mathbf{T} - \rho\mathbf{b}]\, dx$$
$$+ \int_{\Omega} \boldsymbol{\tau} * [\mathrm{sym}\nabla\mathbf{u} - \mathbf{E}_0 - (\mathbf{J}_0\mathbf{T} + \mathbf{J}' * \mathbf{T})]\, dx$$
$$+ \int_{\Omega} \rho[\mathbf{w} \cdot (\dot{\mathbf{u}}(0) - \dot{\mathbf{u}}_0) + \dot{\mathbf{w}} \cdot (\mathbf{u}(0) - \mathbf{u}_0)]\, dx$$
$$+ \int_{\partial\Omega_T} \mathbf{w} * (\mathbf{Tn} - \hat{\mathbf{t}})\, da + \int_{\partial\Omega_u} \boldsymbol{\tau}\,\mathbf{n} * (\hat{\mathbf{u}} - \mathbf{u})\, da.$$

Let $\mathbf{w} = 0$ on $\partial\Omega_T$, $\boldsymbol{\tau} = 0$ on $\partial\Omega_u$ identically in the time interval $[0, t]$ under consideration and $\mathbf{w}(t)$, $\dot{\mathbf{w}}(t) = 0$. Then $d\Upsilon(\zeta \mid \omega) = 0$ for any admissible $\omega \in \tilde{H}$ only if the equation of motion and the creep constitutive relation hold. The arbitrariness and independence of $\mathbf{w}$ on $\partial\Omega_T \times [0, t]$, of $\boldsymbol{\tau}$ on $\partial\Omega_u \times [0, t]$, and of $\mathbf{w}$ and $\dot{\mathbf{w}}$ on Ω at time t provide the boundary conditions and the initial conditions, respectively. This completes the "only if" part. As usual, the proof of the "if" part is obvious. □

It is a simple matter to arrive at the counterpart of Theorem 5.3.3. Letting

$$\tilde{\boldsymbol{\mathcal{F}}}(\mathbf{x}, t) = (g * \rho\mathbf{b})(\mathbf{x}, t) + \rho[t\, \dot{\mathbf{u}}(\mathbf{x}, 0) + \mathbf{u}(\mathbf{x}, 0)],$$

we can write the equations (problem $\tilde{\mathrm{P}}_{\mathrm{c}}$) as

$$\mathrm{sym}\nabla\mathbf{u} - \mathbf{E}_0 - \mathbf{J}_0\mathbf{T} - \mathbf{J}' * \mathbf{T} = 0,$$
$$\rho\mathbf{u} - g * \nabla \cdot \mathbf{T} - \tilde{\boldsymbol{\mathcal{F}}} = 0$$

on $\Omega \times \mathbb{R}^+$ and the boundary conditions

$$\mathbf{u} = \hat{\mathbf{u}} \quad \text{on } \partial\Omega_u \times \mathbb{R}^+, \qquad \mathbf{Tn} = \hat{\mathbf{t}} \quad \text{on } \partial\Omega_T \times \mathbb{R}^+.$$

As with the problem P_{c}, the potentialness condition holds if the constitutive equation is convoluted with g. In this case we apply (5.1.2) and determine the desired functional which involves the convolution of the first equation with $\mathbf{T}$ and that of the second one with $\mathbf{u}$. The addition of appropriate boundary terms leads to the following variational formulation.

THEOREM 5.4.2. *The functional*

$$\Upsilon_0(\zeta) = \int_{\Omega} [g * \mathbf{T} * (\mathrm{sym}\nabla\mathbf{u} - \mathbf{E}_0) - \tfrac{1}{2} g * \mathbf{T} * (\mathbf{J}_0\mathbf{T} + \mathbf{J}' * \mathbf{T})$$
$$+ \tfrac{1}{2}\rho\mathbf{u} * \mathbf{u} - \tilde{\boldsymbol{\mathcal{F}}} * \mathbf{u}]\, dx$$
$$+ \int_{\partial\Omega_u} g * (\hat{\mathbf{u}} - \mathbf{u}) * \mathbf{Tn}\, da - \int_{\partial\Omega_T} g * \hat{\mathbf{t}} * \mathbf{u}\, da$$

on $\tilde{H}$ is stationary at $\zeta \in \tilde{H}$ if and only if ζ is a solution of the mixed problem $\tilde{\mathrm{P}}_c$.

Proof. By proceeding along the usual lines, we consider the vanishing of $d\Upsilon_0(\zeta \mid \omega)$ in $\tilde{H}$ and make use of the arbitrariness and independence of $\boldsymbol{\tau}$ and $\mathbf{w}$ (cf. [92], Thm. 4.2).[31] □

To complete our scheme of variational formulations, we set up a variational formulation involving only stresses, whose case is the formal analogue of that in Theorem 5.2.1. In this regard, consider the problem $\tilde{\mathrm{P}}_c$ and apply the gradient operator to the equation of motion. Upon symmetrization we have

$$\rho\mathbf{E} = g * \mathrm{sym}(\nabla \otimes \nabla \cdot \mathbf{T}) + \mathrm{sym}(\nabla \otimes \tilde{\boldsymbol{\mathcal{F}}}).$$

Substitution of $\mathbf{E}$ as given by the creep law provides

$$g * \mathrm{sym}(\nabla \otimes \nabla \cdot \mathbf{T}) + \mathrm{sym}(\nabla \otimes \tilde{\boldsymbol{\mathcal{F}}}) - \rho\mathbf{E}_0 - \rho(\mathbf{J}_0\,\mathbf{T} + \mathbf{J}' * \mathbf{T}) = 0. \tag{5.4.1}$$

As to the boundary conditions, the knowledge of $\mathbf{u}$ on $\partial\Omega_u$ and the equation of motion amount to setting

$$g * \nabla \cdot \mathbf{T} = \rho\hat{\mathbf{u}} - \tilde{\boldsymbol{\mathcal{F}}} \quad \text{on } \partial\Omega_u \times \mathbb{R}^+. \tag{5.4.2}$$

Moreover, we have

$$\mathbf{T}\,\mathbf{n} = \hat{\mathbf{t}} \quad \text{on } \partial\Omega_T \times \mathbb{R}^+. \tag{5.4.3}$$

No time derivatives are involved and then explicit initial conditions are not required. Equation (5.4.1), along with (5.4.2) and (5.4.3), constitute the formulation of the mixed problem in terms of the symmetric stress only (problem P_s).

Equation (5.4.1) meets the potentialness condition (5.1.1). According to (5.1.2), convolution of (5.4.1) with the symmetric stress $\mathbf{T}$ and integration by parts yield the desired functional. The variational formulation is made precise by the following theorem (cf. [92]).

THEOREM 5.4.3. *Let* $\mathbf{E}_0 : \Omega \times \mathbb{R}^+ \to \mathrm{Sym}$ *be given and* $\Gamma = C^{2,0}(\Omega \times \mathbb{R}^+, \mathrm{Sym})$. *The functional*

$$\begin{aligned}\Sigma(\mathbf{T}) = \tfrac{1}{2} \int_\Omega & [g * \nabla \cdot \mathbf{T} * \nabla \cdot \mathbf{T} + \rho\mathbf{T} * (\mathbf{J}_0\,\mathbf{T} + \mathbf{J}' * \mathbf{T}) \\ & - 2\,\mathbf{T} * (\mathrm{sym}(\nabla \otimes \tilde{\boldsymbol{\mathcal{F}}}) - \rho\mathbf{E}_0)]\,dx \\ & + \int_{\partial\Omega_u} (\tilde{\boldsymbol{\mathcal{F}}} - \rho\hat{\mathbf{u}}) * \mathbf{T}\,\mathbf{n}\,da + \int_{\partial\Omega_T} g * (\hat{\mathbf{t}} - \mathbf{T}\,\mathbf{n}) * \nabla \cdot \mathbf{T}\,da\end{aligned}$$

[31]In connection with [92] it is worth emphasizing that $\mathbf{E}$ or $\tilde{\mathbf{E}}$ is not among the unknown functions. That is why here we write $\mathrm{sym}\nabla\mathbf{u}$ rather than $\mathbf{E}$.

on Γ *is stationary at* $\mathbf{T} \in \Gamma$ *if and only if* $\mathbf{T}$ *is a solution of the mixed problem* P_s.

Proof. Letting $\mathbf{T}, \boldsymbol{\tau} \in \Gamma$ we have

$$
\begin{aligned}
d\Sigma(\mathbf{T} \mid \boldsymbol{\tau}) = \int_\Omega & [g * \nabla \cdot \mathbf{T} * \nabla \cdot \boldsymbol{\tau} + \rho\boldsymbol{\tau} * (\mathbf{J}_0\, \mathbf{T} + \mathbf{J}' * \mathbf{T}) \\
& - \boldsymbol{\tau} * (\operatorname{sym}(\nabla \otimes \tilde{\boldsymbol{\mathcal{F}}}) - \rho \mathbf{E}_0)]\, dx \\
& + \int_{\partial\Omega_u} (\tilde{\boldsymbol{\mathcal{F}}} - \rho\hat{\mathbf{u}}) * \boldsymbol{\tau}\, \mathbf{n}\, da + \int_{\partial\Omega_T} g * [(\hat{\mathbf{t}} - \mathbf{T}\, \mathbf{n}) * \nabla \cdot \boldsymbol{\tau} - \boldsymbol{\tau}\, \mathbf{n} * \nabla \cdot \mathbf{T}]\, da.
\end{aligned}
$$

An integration by parts gives

$$
\begin{aligned}
d\Sigma(\mathbf{T} \mid \boldsymbol{\tau}) = - \int_\Omega & [g * \operatorname{sym}\nabla \otimes \nabla \cdot \mathbf{T} * \boldsymbol{\tau} - \rho\boldsymbol{\tau} * (\mathbf{J}_0\, \mathbf{T} + \mathbf{J}' * \mathbf{T}) \\
& + \boldsymbol{\tau} * (\operatorname{sym}(\nabla \otimes \tilde{\boldsymbol{\mathcal{F}}}) - \rho \mathbf{E}_0)]\, dx \\
& + \int_{\partial\Omega_u} (g * \nabla \cdot \mathbf{T} + \tilde{\boldsymbol{\mathcal{F}}} - \rho\hat{\mathbf{u}}) * \boldsymbol{\tau}\, \mathbf{n}\, da + \int_{\partial\Omega_T} g * (\hat{\mathbf{t}} - \mathbf{T}\, \mathbf{n}) * \nabla \cdot \boldsymbol{\tau}\, da.
\end{aligned}
$$

The arbitrariness of $\boldsymbol{\tau} \in \Gamma$ proves that $d\Sigma(\mathbf{T} \mid \boldsymbol{\tau})$ vanishes only if $\mathbf{T}$ is a solution of the mixed problem P_s. The proof of the "if" part is obvious. □

5.5. Extremum principles. As is the case for the results in the previous sections, usually variational formulations consist in a proper functional being *stationary* at a function in a suitable class, but not necessarily an *extremum.* Really extremum principles are decisively more useful than stationary principles, especially in connection with the development of approximation schemes.

Concerning linear elastodynamics, a minimum principle was established by Benthien and Gurtin [3] for the mixed boundary value problem; they showed that the Laplace transform of the energy functional is minimum at the solution. An improvement is due to Reiss [110] who elaborated a method whereby the Benthien-Gurtin functional can be transformed back to the original space-time domain while preserving the extremum property of the functional. An extension of Reiss' method to viscoelasticity was developed in [17].

Standard variational formulations may become extremum principles upon a proper choice of the class of functions where the functionals are considered. This idea was made operative by Christensen ([16], §5.5) who showed that a minimum principle in viscoelasticity holds in the class of displacement fields for which the time dependence is factorized and not varied. In doing this, he confined the attention to relaxation functions which satisfy thermodynamic requirements in the form of the dissipation principle.

By paralleling [49], in the next sections we investigate some extremum (minimum) properties and show how the extremum property is related to the class of functions (displacements) under consideration. Meanwhile we emphasize that the validity of the extremum properties is crucially related to the thermodynamic restrictions on the relaxation function.

First, though, it is worth introducing some pertinent notation and definitions. By a kinematically admissible displacement field we mean a vector field on $\Omega \times \mathbb{R}^+$ that satisfies the displacement boundary condition $(5.2.3)_1$. Denote by $\mathcal{U}$ the set of all kinematically admissible displacements. We denote by $\mathcal{U}_L$ the set of all $\mathbf{u} \in \mathcal{U}$ that possess Laplace transform $\mathbf{u}_L$ with respect to time in the complex half plane $\mathbb{C}^{++}$. We assume that $\mathbf{T}_0, \mathbf{b}, \hat{\mathbf{u}}, \hat{\mathbf{t}}$ possess Laplace transform in $\mathbb{C}^{++}$. Of course $\mathbf{u}_L$ must comply with

$$\mathbf{u}_L(\mathbf{x}, p) = \hat{\mathbf{u}}_L(\mathbf{x}, p) \quad \text{on } \partial\Omega_u \times \mathbb{C}^{++}. \tag{5.5.1}$$

In connection with the problem P, application of the Laplace transform operator to (5.2.1) and account of (5.2.4) give

$$L_p \mathbf{u}_L = \rho p^2 \mathbf{u}_L - \nabla \cdot [(\mathbf{G}_0 + \mathbf{G}'_L)\nabla \mathbf{u}_L] = \mathbf{g}, \tag{5.5.2}$$

where

$$\mathbf{g} = \rho \mathbf{b}_L + \nabla \cdot (\mathbf{T}_0)_L + \rho(p\mathbf{u}_0 - \dot{\mathbf{u}}_0). \tag{5.5.3}$$

The counterpart of the problem P in the Laplace transform context, say P_L, is then given by

$$L_p \mathbf{u} = \mathbf{g} \quad \text{on } \Omega \times \mathbb{C}^{++} \tag{5.5.4}$$

along with the boundary conditions (5.5.1) and

$$[(\mathbf{G}_0 + \mathbf{G}'_L)\nabla \mathbf{u}_L]\mathbf{n} = (\hat{\mathbf{t}}_0)_L \quad \text{on } \partial\Omega_T \times \mathbb{C}^{++}, \tag{5.5.5}$$

where $\hat{\mathbf{t}}_0 = \hat{\mathbf{t}} - \mathbf{T}_0\mathbf{n}$.

5.6. Minimum principle via weight function. By generalizing a procedure which traces back to Reiss [110], we establish a minimum principle in terms of an appropriate weight function. For this purpose, it is convenient to consider some function spaces as follows.

For the time being let H be any Hilbert space and define L^2_L as

$$L^2_L(\mathbb{R}^+; H) = \{\mathbf{u} \in L^2_{loc}(\mathbb{R}^+, H) : \mathbf{u}_L(p) \in H, \; p \in \mathbb{C}^{++}\}.$$

Moreover, let

$$H^1_L(\mathbb{R}^+; L^2(\Omega)) = \{\mathbf{u} \in H^1_{loc}(\mathbb{R}^+; L^2(\Omega)) : \mathbf{u}_L(p) \in L^2(\Omega), \; p \in \mathbb{C}^{++}\}.$$

Consider a function $Y \in C(\mathbb{R}^+, \mathbb{R}^+)$ such that

$$y(t) := \int_0^\infty Y(s) \exp(-st) ds$$

exists for any $t \in \mathbb{R}^+$. In terms of y we define the bilinear form $(\cdot, \cdot)_y$ on $L^2_L(\mathbb{R}^+; L^2(\Omega))$ as

$$(v, w)_y = \int_0^\infty \int_0^\infty y(t+\tau) \int_\Omega v(\mathbf{x}, t) w(\mathbf{x}, \tau) \, dx \, d\tau \, dt;$$

if (both) v and w are vectors or tensors, then the product vw is meant as the corresponding inner product. By the definition of y we have

$$(v, w)_y = \int_0^\infty \int_\Omega Y(s) v_L(\mathbf{x}, s) w_L(\mathbf{x}, s) \, dx \, ds. \tag{5.6.1}$$

Since $Y \geq 0$, this shows that the bilinear form $(\cdot, \cdot)_y$ is positive semidefinite. Of course $(\cdot, \cdot)_y$ is well defined on L^2_L when $Y \in C_0^\infty$. As before, to save writing, the dependence on $\mathbf{x} \in \Omega$ is understood and not written.

The following property, which is a consequence of the definition of the bilinear form $(\cdot, \cdot)_y$, is helpful when variational formulations are considered which involve convolutions. By definition,

$$\int_0^\infty y(t+\tau)(v * w)(\tau) \, d\tau = \int_0^\infty y(t+\tau) \int_0^\tau v(\tau - \sigma) \, w(\sigma) \, d\sigma \, d\tau.$$

The change of variables $\lambda = \tau - \sigma, \quad \eta = \sigma$ yields

$$\int_0^\infty y(t+\tau) \, (v * w)(\tau) \, d\tau = \int_0^\infty \int_0^\infty y(t + \lambda + \eta) \, v(\lambda) \, w(\eta) \, d\lambda \, d\eta.$$

As a consequence, we find that

$$(u, v * w)_y = (v, u * w)_y = (v, w * u)_y. \tag{5.6.2}$$

We now show that the mixed problem P can be given a variational formulation in terms of $(\cdot, \cdot)_y$. By analogy with § 5.2, consider the functional

$$\begin{aligned}\Lambda_y(\mathbf{u}) = {} & \tfrac{1}{2}(\nabla \mathbf{u}, \mathbf{G}_0 \nabla \mathbf{u} + \mathbf{G}' * \nabla \mathbf{u})_y + \tfrac{1}{2}(\rho \dot{\mathbf{u}}, \dot{\mathbf{u}})_y - (\rho \mathbf{u}, \mathbf{f})_y \\ & - \int_0^\infty \int_0^\infty \int_{\partial \Omega_T} y(t+\tau) \mathbf{u}(t) \cdot \hat{\mathbf{t}}_0(\tau) \, d\tau \, dt \, da \\ & - \int_0^\infty y(t) \int_\Omega \rho \{ \dot{\mathbf{u}}(t) \cdot [\mathbf{u}(0) - \mathbf{u}_0] - \mathbf{u}(t) \cdot \dot{\mathbf{u}}_0 \} \, dx \, dt \\ & + y(0) \int_\Omega \rho \mathbf{u}(0) \cdot [\tfrac{1}{2} \mathbf{u}(0) - \mathbf{u}_0] \, dx\end{aligned}$$

where $\rho\mathbf{f} = \rho\mathbf{b} + \nabla\cdot\mathbf{T}_0$ and $\hat{\mathbf{t}}_0 = \hat{\mathbf{t}} - \mathbf{T}_0\mathbf{n}$ are supposed to belong to $L^2_L(\mathbb{R}^+; L^2(\Omega))$ and $L^2_L(\mathbb{R}^+; L^2(\partial\Omega))$, respectively. Then, in terms of the spaces

$$\mathcal{P} = \{\mathbf{u} \in L^2_L(\mathbb{R}^+; H^1(\Omega)) \cap H^1_L(\mathbb{R}^+; L^2(\Omega)) : \mathbf{u}\,|_{\partial\Omega_u} = \hat{\mathbf{u}}\},$$

$$\mathcal{P}_0 = \{\mathbf{h} \in L^2_L(\mathbb{R}^+; H^1(\Omega)) \cap H^1_L(\mathbb{R}^+; L^2(\Omega)) : \mathbf{h}\,|_{\partial\Omega_u} = 0\},$$

we can establish a minimum principle as follows.

THEOREM 5.6.1. *Let* $\mathbf{G}$ *satisfy the inequality* (3.4.4). *If* $\mathbf{u}^0 \in \mathcal{P}$ *is a solution to the mixed problem* P *with* $\mathbf{f} \in L^2_L(\mathbb{R}^+; L^2(\Omega))$ *and* $\hat{\mathbf{t}}_0 \in L^2_L(\mathbb{R}^+; L^2(\partial\Omega))$, *then, for every function* $Y \in C_0^\infty(\mathbb{R}^+, \mathbb{R}^+)$, *the functional* $\Lambda_y(\mathbf{u})$ *has a minimum at* $\mathbf{u}^0$ *on* $\mathcal{P}$.

Proof. Let $\mathbf{u}^0 \in \mathcal{P}$ and represent any element $\mathbf{u} \in \mathcal{P}$ as $\mathbf{u}^0 + \mathbf{h}$, where $\mathbf{h} \in \mathcal{P}_0$. Then by (5.6.1) we have

$$\begin{aligned}
&\Lambda_y(\mathbf{u}^0 + \mathbf{h}) - \Lambda_y(\mathbf{u}^0) = (\nabla\mathbf{h}, \mathbf{G}_0\nabla\mathbf{u}^0 + \mathbf{G}' * \nabla\mathbf{u}^0)_y + \tfrac{1}{2}(\nabla\mathbf{h}, \mathbf{G}_0\nabla\mathbf{h} + \mathbf{G}' * \nabla\mathbf{h})_y \\
&+ (\rho\dot{\mathbf{h}}, \dot{\mathbf{u}}^0)_y + \tfrac{1}{2}(\rho\dot{\mathbf{h}}, \dot{\mathbf{h}})_y - (\rho\mathbf{h}, \mathbf{f})_y \\
&- \int_0^\infty \int_0^\infty y(t+\tau) \int_{\partial\Omega_T} \mathbf{h}(t)\cdot\hat{\mathbf{t}}_0(\tau)\, da\, d\tau\, dt \\
&- \int_0^\infty y(t) \int_\Omega \rho\{[\dot{\mathbf{u}}^0(t) + \dot{\mathbf{h}}(t)]\cdot\mathbf{h}(0) + [\mathbf{u}^0(0) - \mathbf{u}_0]\cdot\dot{\mathbf{h}}(t) - \mathbf{h}(t)\cdot\dot{\mathbf{u}}_0\}\, dx\, dt \\
&+ \tfrac{1}{2}y(0) \int_\Omega \rho\mathbf{h}(0)\cdot\{\mathbf{h}(0) + 2[\mathbf{u}^0(0) - \mathbf{u}_0]\}\, dx.
\end{aligned}$$

Suitable integrations by parts and the fact that $\mathbf{u}^0$ is a solution to the mixed problem P lead to

$$\begin{aligned}
\Lambda_y(\mathbf{u}^0 + \mathbf{h}) - \Lambda_y(\mathbf{u}^0) &= \tfrac{1}{2}(\nabla\mathbf{h}, \mathbf{G}_0\nabla\mathbf{h} + \mathbf{G}' * \nabla\mathbf{h})_y + \tfrac{1}{2}(\rho\dot{\mathbf{h}}, \dot{\mathbf{h}})_y \\
&- \int_0^\infty y(t) \int_\Omega \rho\dot{\mathbf{h}}(t)\cdot\mathbf{h}(0)\, dx\, dt + \tfrac{1}{2}\rho y(0) \int_\Omega \mathbf{h}(0)\cdot\mathbf{h}(0)\, dx.
\end{aligned}$$

Hence, by (5.6.2) we have

$$\begin{aligned}
&\Lambda_y(\mathbf{u}^0 + \mathbf{h}) - \Lambda_y(\mathbf{u}^0) \\
&= \tfrac{1}{2}\int_\Omega \int_0^\infty Y(s)[\nabla\mathbf{h}_L(s)\cdot(\mathbf{G}_0 + \mathbf{G}'_L(s))\nabla\mathbf{h}_L(s) + \rho s^2\mathbf{h}_L(s)\cdot\mathbf{h}_L(s)]\, ds\, dx.
\end{aligned}$$

By (3.4.4), $\mathbf{G}_0 + \mathbf{G}'_L(s)$ is positive definite for any $s \in \mathbb{R}^+$ and this makes $\Lambda_y(\mathbf{u}^0 + \mathbf{h}) \geq \Lambda_y(\mathbf{u}^0)$ for any $\mathbf{h} \in \mathcal{P}_0$. □

The converse of Theorem 5.6.1 needs a stronger condition on $\mathbf{G}'$, namely the strict inequality (3.2.11), rather than (3.4.1), and (3.1.1).

THEOREM 5.6.2. *Let* $\mathbf{G}$ *satisfy the inequalities* (3.1.1) *and* (3.2.11). *If the function* $\mathbf{w} \in \mathcal{P}$ *minimizes the functional* Λ_y *on* $\mathcal{P}$, *for every* $Y \in C_0^\infty(\mathbb{R}^+, \mathbb{R}^+)$, *then* $\mathbf{w}$ *is the unique solution to the problem* P *with* $\mathbf{f} \in L_L^2(\mathbb{R}^+; L^2(\Omega))$ *and* $\hat{\mathbf{t}}_0 \in L_L^2(\mathbb{R}^+; L^2(\partial\Omega))$.

Proof. Let $\mathbf{w} \in \mathcal{P}$ be a function which minimizes Λ_y in $\mathcal{P}$ for every $Y \in C_0^\infty(\mathbb{R}^+)$. Then for every $\mathbf{h} \in \mathcal{P}_0$ it follows that

$$\begin{aligned} 0 = d\Lambda_y(\mathbf{w} \mid \mathbf{h}) &= (\nabla\mathbf{h}, \mathbf{G}_0\nabla\mathbf{w} + \mathbf{G}' * \nabla\mathbf{w})_y + (\rho\dot{\mathbf{h}}, \dot{\mathbf{w}})_y - (\rho\mathbf{h}, \mathbf{f})_y \\ &- \int_0^\infty \int_0^\infty y(t+\tau) \int_{\partial\Omega_T} \hat{\mathbf{t}}_0(\tau) \cdot \mathbf{h}(t) da\, d\tau\, dt \\ &- \int_0^\infty y(t) \int_\Omega \rho\{\dot{\mathbf{h}}(t) \cdot [\mathbf{w}(0) - \mathbf{u}_0] + \mathbf{h}(0) \cdot \dot{\mathbf{w}}(t) - \mathbf{h}(t) \cdot \dot{\mathbf{u}}_0\} dx\, dt \\ &+ y(0)\rho \int_\Omega [\mathbf{w}(0) - \mathbf{u}_0] \cdot \mathbf{h}(0) dx. \end{aligned}$$

Let $\tilde{\boldsymbol{T}}$ denote the stress functional given by $\tilde{\boldsymbol{T}}(\nabla\mathbf{u}^t) = (\mathbf{G}_0\nabla\mathbf{u} + \mathbf{G}' * \nabla\mathbf{u})(t)$. Upon suitable integrations by parts, we obtain

$$\begin{aligned} &\int_0^\infty \int_0^\infty y(t+\tau) \bigg\{ \int_\Omega [\rho\ddot{\mathbf{w}}(t) - \nabla \cdot \tilde{\boldsymbol{T}}(\nabla\mathbf{w}^t) - \rho\mathbf{f}(t)] \cdot \mathbf{h}(\tau) dx \\ &\qquad + \int_{\partial\Omega_T} [\tilde{\boldsymbol{T}}(\nabla\mathbf{w}^t)\mathbf{n} - \tilde{\mathbf{t}}_0(t)] \cdot \mathbf{h}(\tau) da \bigg\} d\tau\, dt \\ &- \int_0^\infty y(t) \int_\Omega \rho\{[\dot{\mathbf{w}}(0) - \dot{\mathbf{u}}_0] \cdot \mathbf{h}(t) + [\mathbf{w}(0) - \mathbf{u}_0] \cdot \dot{\mathbf{h}}(t)\} dx\, dt \\ &+ y(0) \int_\Omega \rho[\mathbf{w}(0) - \mathbf{u}_0] \cdot \mathbf{h}(0) dx = 0 \end{aligned}$$

whence, by (5.6.1), we have

$$\begin{aligned} &\int_0^\infty Y(s) \bigg\{ \int_\Omega [L_s\mathbf{w}_L(s) - \mathbf{g}(s)] \cdot \mathbf{h}_L(s)\, dx \\ &\qquad + \int_{\partial\Omega_T} \{[(\mathbf{G}_0 + \mathbf{G}'_L(s))\nabla\mathbf{w}_L(s)]\mathbf{n} - (\hat{\mathbf{t}}_0)_L(s)\} \cdot \mathbf{h}_L(s)\, da \bigg\} ds = 0 \end{aligned} \tag{5.6.3}$$

where L_s and $\mathbf{g}$ are the operator and the function defined by (5.5.2) and (5.5.3). Let $\mathbf{h}(\mathbf{x}, t)$ take the form

$$\mathbf{h}(\mathbf{x}, t) = \mathbf{h}_1(\mathbf{x}) h_2(t)$$

where $\mathbf{h}_1 \in H_0^1(\Omega), h_2 \in H^1(\mathbb{R}^+)$. By the arbitrariness of $\mathbf{h}_1$ we obtain

$$\int_0^\infty Y(s)[L_s \mathbf{w}_L(s) - \mathbf{g}(s)](h_2)_L(s) ds = 0$$

almost everywhere in Ω. Hence the arbitrariness of $Y(h_2)_L$ on $\mathbb{R}^+$ gives

$$L_s \mathbf{w}_L(s) - \mathbf{g}(s) = 0 \tag{5.6.4}$$

almost everywhere in $\Omega \times \mathbb{R}^+$. The arbitrariness of $Y\mathbf{h}_L$ on $\partial\Omega_T \times \mathbb{R}^+$ and (5.6.4) make (5.6.3) reduce to

$$[(\mathbf{G}_0 + \mathbf{G}'_L(s))\nabla \mathbf{w}(s)]\mathbf{n} = (\hat{\mathbf{t}}_0)_L(s) \tag{5.6.5}$$

almost everywhere in $\partial\Omega_T \times \mathbb{R}^+$. Accordingly, $\mathbf{w}_L$ is a solution to the problem P_L (see (5.5.4), (5.5.5)) for almost every $s \in \mathbb{R}^+$.

As shown in § 4.6, the conditions (3.1.1) and (3.2.11) provide the invertibility of $\mathbf{G}_0 + \mathbf{G}'_L(p)$ in Sym for any $p \in \mathbb{C}^+$. Then the operator

$$L_p = \rho p^2 - \nabla \cdot [(\mathbf{G}_0 + \mathbf{G}'_L(p))\nabla], \qquad p \in \mathbb{C}^+,$$

is strongly elliptic in $H^1(\Omega)$. As a consequence, $\mathbf{w}_L$ is the unique solution to P_L. Now, let $\mathbf{v}$ be the solution to the problem P which exists and is unique in $\mathcal{P}$. Its Laplace transform $\mathbf{v}_L$ exists for any $p \in \mathbb{C}^{++}$ and is a solution to the problem P_L. This implies that $\mathbf{v}_L(s)$ and $\mathbf{w}_L(p)$ coincide when $p = s$. Since the Laplace transform $\mathbf{w}_L(p)$ of the minimizing function $\mathbf{w}$ and $\mathbf{v}_L(p)$ are analytic functions which coincide on $\mathbb{R}^{++}$, then they must coincide in the complex half plane $\mathbb{C}^{++}$. Hence, by the uniqueness of the Laplace transform, the minimizing function $\mathbf{w}$ is the unique solution to the problem P in $\mathcal{P}$. □

5.7. Minimum principle via Laplace transform. The solution $\mathbf{w}^0(p)$ to the problem P_L is obviously a stationary point of the functional

$$\begin{aligned}\mathcal{L}_p(\mathbf{w}) = \int_\Omega \{\tfrac{1}{2}\rho p^2 \mathbf{w}(p)\cdot\mathbf{w}(p) + \tfrac{1}{2}\nabla\mathbf{w}(p)\cdot[\mathbf{G}_0 + \mathbf{G}'_L(p)]\nabla\mathbf{w}(p) - \mathbf{g}(p)\cdot\mathbf{w}(p)\}dx \\ - \int_{\partial\Omega_T} (\hat{\mathbf{t}}_0)_L(p)\cdot\mathbf{w}(p)\, da\end{aligned} \tag{5.7.1}$$

which is parameterized by the complex quantity p both explicitly and through $\mathbf{w}(p)$. Letting

$$\Gamma_p(\mathbf{u}) = \mathcal{L}_p(\mathbf{u}_L(p)) \tag{5.7.2}$$

we can regard Γ_p as a functional on $\mathcal{U}_L$ parameterized by the complex quantity p. Indeed, a solution $\mathbf{u}^0 \in \mathcal{U}_L$ to the problem P_L is a stationary point of Γ_p for any $p \in \mathbb{C}^{++}$. The next theorem shows how the stationary point is indeed a minimum of the functional Γ_p.

THEOREM 5.7.1. *Let* $\mathbf{G}$ *satisfy the inequality* (3.4.4). *If a function* $\mathbf{u}^0 \in \mathcal{U}_L$ *is a solution to the problem* P, *then, for any* $\alpha \in \mathbb{R}^+$, *the functional* Γ_α *on* $\mathcal{U}_L$ *has a strict minimum at* $\mathbf{u}^0$.

Proof. Let $\mathbf{u}^0 \in \mathcal{U}_L$ be a solution to P. Then $\mathbf{u}_L^0(p)$, $p \in \mathbb{C}^{++}$, satisfies (5.5.1), (5.5.4), (5.5.5) and is a stationary point of (5.7.1). Moreover, letting $\mathbf{u}^0 + \nu\mathbf{h} \in \mathcal{U}_L$ as $\nu \in [0, \bar{\nu}]$, $\bar{\nu} \in \mathbb{R}^{++}$, we have

$$d^2\mathcal{L}_p(\mathbf{u}_L^0(p)|\mathbf{h}_L(p)) = \int_\Omega [\rho p^2 \mathbf{h}_L(p)\cdot\mathbf{h}_L(p) + \nabla\mathbf{h}_L(p)\cdot(\mathbf{G}_0 + \mathbf{G}_L'(p))\nabla\mathbf{h}_L(p)]dx.$$

Let p be real, that is, $p = \alpha \in \mathbb{R}^{++}$. Then $\mathbf{h}_L(\alpha), \nabla\mathbf{h}_L(\alpha)$, and $\mathbf{G}_L'(\alpha)$ are real-valued as well as $\mathcal{L}_\alpha$. Meanwhile, by (3.4.4), we have $\mathbf{G}_0 + \tilde{\mathbf{G}}'(\alpha) > 0$ and hence, for any nonzero admissible $\mathbf{h}_L$, it follows that

$$d^2\mathcal{L}_\alpha(\mathbf{u}_L^0(\alpha)|\mathbf{h}_L(\alpha)) > 0.$$

The vanishing of $\mathbf{h}_L(\alpha)$ on $\mathbb{R}^{++}$ implies the vanishing of $\mathbf{h}$. Then by (5.7.2) it follows that

$$d^2\Gamma_\alpha(\mathbf{u}^0|\mathbf{h}) > 0$$

for any nonzero admissible $\mathbf{h}$ and for any $\alpha \in \mathbb{R}^{++}$, which completes the proof. □

Remark 5.7.1. As shown in Chapter 3, for a viscoelastic body the inequality (3.4.4) is a consequence of (3.1.1) and the thermodynamic restriction (3.4.1). The converse does not hold. Thus the hypothesis in Theorem 5.7.1 is weaker than the thermodynamic restrictions.

It is natural to ask whether the converse property holds, namely that if there exists $\mathbf{u}^0 \in \mathcal{U}_L$ which is a strict minimum of Γ_α, $\alpha \in \mathbb{R}^+$, then $\mathbf{u}^0$ is the solution to the problem P. This property is in fact nontrivial because the solution to P_L is a function of the complex variable p. The property is shown to hold by assuming the stronger condition (3.2.11) in place of (3.4.1).

THEOREM 5.7.2. *Let* $\mathcal{Q} = \{\mathbf{u} \in \mathcal{U}_L : \mathbf{u}_L(p) \in H^1(\Omega), p \in \mathbb{C}^{++}\}$ *and* $\mathbf{G}$ *satisfy the conditions* (3.1.1) *and* (3.2.11). *If* Γ_α, $\alpha \in \mathbb{R}^{++}$, *has a strict minimum in* $\mathcal{Q}$ *at* $\mathbf{u}^0$, *then* $\mathbf{u}^0$ *is a solution to the problem* P.

Proof. Since Γ_α has a strict minimum at $\mathbf{u}^0$ then by (5.7.2) $\mathcal{L}_\alpha$ has a strict minimum at $\mathbf{u}_L^0(\alpha)$. This means that, for any $\alpha \in \mathbb{R}^{++}$,

$$\begin{aligned} 0 &= d\mathcal{L}_\alpha(\mathbf{u}_L^0(\alpha)|\mathbf{h}_L(\alpha)) \\ &= \int_\Omega \{\rho\alpha^2\mathbf{u}_L^0(\alpha) - \nabla\cdot[(\mathbf{G}_0 + \mathbf{G}_L'(\alpha))\nabla\mathbf{u}_L^0(\alpha)] - \mathbf{g}(\alpha)\}\cdot\mathbf{h}_L(\alpha)dx \\ &\quad + \int_{\partial\Omega_T} \{[(\mathbf{G}_0 + \mathbf{G}_L'(\alpha))\nabla\mathbf{u}_L^0(\alpha)]\mathbf{n} - (\hat{\mathbf{t}}_0)_L(\alpha)\}\cdot\mathbf{h}_L(\alpha)da, \end{aligned} \tag{5.7.3}$$

for any admissible $\mathbf{h}_L$, and

$$
\begin{aligned}
0 <& d^2\mathcal{L}_\alpha(\mathbf{u}_L^0(\alpha)|\mathbf{h}_L(\alpha)) \\
=& \int_\Omega \{\rho\alpha^2\mathbf{h}_L(\alpha)\cdot\mathbf{h}_L(\alpha) + \nabla\mathbf{h}_L(\alpha)\cdot[\mathbf{G}_0 + \mathbf{G}'_L(\alpha)]\nabla\mathbf{h}_L(\alpha)\}dx
\end{aligned}
\tag{5.7.4}
$$

for any nonzero admissible $\mathbf{h}_L(\alpha)$. Then, because of (5.7.3), for any $\alpha \in \mathbb{R}^+$ the function $\mathbf{u}_L^0(\alpha)$ satisfies (5.5.4) and (5.5.5), i.e., it is a solution to the problem P_L for real values of p. We know that (3.1.1) and (3.4.1) imply (3.4.2) and (3.4.4). Now, letting $p = \alpha + i\beta$, we have

$$
\begin{aligned}
\mathbf{G}_0 + \mathbf{G}'_L(p) = \mathbf{G}_0 &+ \int_0^\infty \mathbf{G}'(s)\exp(-\alpha s)\cos\beta s\, ds \\
&+ i\int_0^\infty \mathbf{G}'(s)\exp(-\alpha s)\sin\beta s\, ds.
\end{aligned}
$$

Then, by (3.4.2) and (3.4.4), the complex operator L_p turns out to be strongly elliptic for any $p \in \mathbb{C}^{++}$. Hence the problem P_L has a unique solution $\mathbf{w}^0(p) \in H^1(\Omega)$ which is analytic in p on $\mathbb{C}^{++}$ and equal to $\mathbf{u}_L^0$ on the real positive half line. Then it follows that $\mathbf{w}^0$ and $\mathbf{u}_L^0$ coincide on the whole half space $\mathbb{C}^{++}$. Finally, the uniqueness of the Laplace transform implies that $\mathbf{u}^0$ is a solution to the problem P. □

5.8. Extremum principle and stationary principle for the quasi-static problem. While being a nondegenerate form, the convolution is not a definite form in that the sign of $\mathbf{u} * \mathbf{u}$, say, depends on $\mathbf{u}$ itself. Then there is no possibility of getting extremum principles by considering the variational formulations of §§5.2–5.4. In the meantime this observation suggests that we search for extremum principles via the L^2 inner product as the bilinear form. However, since a convolution inescapably occurs through the stress-strain relation, an extremum property should involve appropriate hypotheses and choices.

On the one hand, an answer to this problem has been given in the previous section. An alternative, though more restrictive, answer is as follows. For strain tensor functions $\mathbf{E}(t)$ that vanish as $t \to -\infty$, we can write the stress-strain relation as (cf. (3.1.3))

$$
\mathbf{T} = \mathbf{G} \circ \dot{\mathbf{E}} := \int_0^\infty \mathbf{G}(s)\mathbf{E}(t-s)\, ds.
$$

Then, in the quasi-static approximation, where the inertia $\rho\ddot{\mathbf{u}}$ is disregarded, the equation of motion (rather, equilibrium) can be written as

$$
\nabla\cdot(\mathbf{G}\circ\nabla\dot{\mathbf{u}}) + \rho\,\mathbf{b} = 0; \tag{5.8.1}
$$

to (5.8.1) we associate the boundary conditions of the mixed problem P. Equation (5.8.1) does not admit a variational formulation in $L^2(\Omega \times \mathbb{R})$ merely because, in general,

$$\langle \nabla \mathbf{u}_1, \mathbf{G} \circ \nabla \dot{\mathbf{u}}_2 \rangle \neq \langle \nabla \mathbf{u}_2, \mathbf{G} \circ \nabla \dot{\mathbf{u}}_1 \rangle$$

and the same happens if the first argument is replaced with its time derivative. This lack of symmetry is due to the time dependence of $\mathbf{u}(\mathbf{x}, t)$ on t which is involved in both the inner product and the convolution. Symmetry is recovered if the time dependence is *factorized* and *not varied.* Accordingly, following [16], §5.5, we let

$$\mathbf{u}(\mathbf{x}, t) = [\mathbf{u}^0(\mathbf{x}) + \mathbf{w}(\mathbf{x})]h(t) \tag{5.8.2}$$

where $\mathbf{u}^0, \mathbf{w} \in C^2(\Omega, V)$, $h \in C^2(\mathbb{R}, \mathbb{R})$, and $\mathbf{u}^0 h$ is a solution to the mixed problem P. We denote by $\mathcal{U}'$ the set of (kinematically admissible) displacement fields characterized by (5.8.2).

Letting t be the present time, consider the functional

$$\Phi(\mathbf{u}) = \int_{-\infty}^{t} \int_{\Omega} (\tfrac{1}{2} \nabla \dot{\mathbf{u}} \cdot \mathbf{G} * \nabla \dot{\mathbf{u}} - \rho \mathbf{b} \cdot \dot{\mathbf{u}}) \, dx \, d\tau - \int_{-\infty}^{t} \int_{\partial \Omega_T} \hat{\mathbf{t}} \cdot \dot{\mathbf{u}} \, da \, d\tau.$$

The functional $\Phi(\mathbf{u})$ is endowed with the following minimum property induced by the dissipation principle (cf. § 2.6).

THEOREM 5.8.1. *If* $\mathbf{G}$ *meets the dissipation principle, then the functional* $\Phi(\mathbf{u})$ *has a minimum in* $\mathcal{U}'$ *at any solution* $\mathbf{u}^0(\mathbf{x})h(t)$ *to* (5.8.1) *with the boundary conditions of* P.

Proof. Upon substitution from (5.8.1) we have

$$\begin{aligned} \Phi(\mathbf{u}) = \int_{\Omega} (\nabla \mathbf{u}^0 + \nabla \mathbf{w}) \cdot \int_{-\infty}^{t} \int_{-\infty}^{\tau} \dot{h}(\tau) \mathbf{G}(\tau - \sigma) \dot{h}(\sigma) \, d\sigma \, d\tau \, (\nabla \mathbf{u}^0 + \nabla \mathbf{w}) \, dx \\ - \int_{\Omega} \rho (\mathbf{u}^0 + \mathbf{w}) \cdot \int_{-\infty}^{t} \mathbf{f}(\tau) \dot{h}(\tau) \, d\tau \, dx \\ - \int_{\partial \Omega_T} (\mathbf{u}^0 + \mathbf{w}) \cdot \int_{-\infty}^{t} \hat{\mathbf{t}}(\tau) \dot{h}(\tau) \, d\tau \, da. \end{aligned}$$

Then on evaluating $d\Phi(\mathbf{u}^0 h \mid \mathbf{w} h)$ and making an integration by parts over $\mathbf{x}$ we see at once that $d\Phi(\mathbf{u}^0 h \mid \mathbf{w} h)$ vanishes at any solution to (5.8.1) with $\mathbf{w} = 0$ on $\partial \Omega_u$ and $\mathbf{T}\mathbf{n} = \hat{\mathbf{t}}$ on $\partial \Omega_T$. Next we have

$$d^2 \Phi(\mathbf{u}^0 h \mid \mathbf{w} h) = \int_{\Omega} \nabla \mathbf{w} \cdot \left[\int_{-\infty}^{t} \int_{-\infty}^{\tau} \dot{h}(\tau) \mathbf{G}(\tau - \sigma) \dot{h}(\sigma) \, d\sigma \, d\tau \right] \nabla \mathbf{w} \, dx.$$

The conclusion follows by observing that the dissipation principle implies the positive definiteness of the tensor in brackets for any $h \in C^2(\mathbb{R}, \mathbb{R})$. □

We append a uniqueness result about the solution to the quasi-static problem. Assume, for simplicity, homogeneous boundary conditions as

$$\partial\Omega_T = \emptyset, \qquad \hat{\mathbf{u}} = 0 \quad \text{on } \partial\Omega \times \mathbb{R}.$$

Then we have the following theorem about fields in $\mathcal{U}'$.

THEOREM 5.8.2. *Let* $\mathbf{G}$ *satisfy the conditions* (3.1.1), (3.2.11). *If* $\mathbf{b} \in W^{1,1}(\mathbb{R}; L^2(\Omega))$ *then the quasi-static problem has a unique solution* $\mathbf{u}^0 h$ *such that* $\mathbf{u}^0 \in H_0^1(\Omega)$ *and* $h \in W^{1,1}(\mathbb{R})$.

Proof. By (3.1.1), (3.2.11), and the homogeneous boundary conditions, it follows from [50] that a unique solution exists in $L^1(\mathbb{R}; H_0^1(\Omega))$. Moreover $\mathbf{u}^0 \dot{h}$ belongs to the same space in that it satisfies the equation obtained by differentiating $\nabla \cdot \mathbf{T} + \rho\mathbf{b} = 0$. Then the desired result follows. □

A more sophisticated way to circumvent the lack of symmetry in the bilinear form $\langle \nabla\mathbf{u}_1, \mathbf{G} * \nabla\dot{\mathbf{u}}_2 \rangle$ is here exhibited.[32] Let $Y \in L^1(\mathbb{R}^+)$ and Υ be the odd extension of Y, namely

$$\Upsilon(s) = \begin{cases} Y(s), & s \in \mathbb{R}^{++}, \\ -Y(s), & s \in \mathbb{R}^{--}. \end{cases}$$

Then define υ on $\mathbb{R}$ as

$$\upsilon(t) = -i \int_{-\infty}^{\infty} \Upsilon(u) \exp(-iut)\, du,$$

that is, $\upsilon = -i\Upsilon_F$. Since Υ is odd we have

$$\upsilon = -2Y_s.$$

It follows easily that $\upsilon(t)$ is a real, odd, absolutely continuous function which vanishes as $|t|$ tends to infinity.

For any pair of scalar-, vector- or tensor-valued functions u, w on $\Omega \times \mathbb{R}$, consider the bilinear form

$$\langle u, w \rangle_\upsilon := \int_{-\infty}^{\infty} \int_{-\infty}^{\infty} \upsilon(t+\tau) \int_\Omega u(t) w(\tau)\, dx\, d\tau\, dt.$$

Substitution for υ yields

$$\langle u, w \rangle_\upsilon = -i \int_{-\infty}^{\infty} \Upsilon(\omega) \int_\Omega u_F(\omega) w_F(\omega)\, dx\, d\omega.$$

---

[32]Here we follow a recent work by Giorgi and Marzocchi.

Since Υ is an odd function and $\Re(u_F\, w_F)$ an even one, then

$$\langle u, w\rangle_v = \int_{-\infty}^{\infty} \Upsilon(\omega)\Im(u_F, w_F)d\omega \tag{5.8.3}$$

where $(\cdot,\cdot)$ is the $L^2(\Omega)$ inner product. It is apparent from (5.8.3) that $\langle\cdot,\cdot\rangle_v$ is symmetric. Moreover, since $u_F = u_C - iu_S$ we can write

$$\langle u, w\rangle_v = -2\int_0^{\infty} \Upsilon[(u_C(\omega), w_S(\omega)) + (u_S(\omega), w_C(\omega))]\, d\omega.$$

For later convenience we observe that, by definition,

$$\langle u \circ v, w\rangle_v = \int_{-\infty}^{\infty} v(t+\tau)\Big(\int_{-\infty}^{\infty} u(\tau-\xi)\, v(\xi)\, d\xi,\ w(t)\Big)\, d\tau\, dt.$$

Upon the change of variables $\lambda = \tau - \xi$, $\nu = \xi$ and (5.8.3), we obtain,

$$\langle u \circ v, w\rangle_v = \int_{-\infty}^{\infty} \Upsilon(\omega)\, \Im(u_F(\omega)\, v_F(\omega), w_F(\omega))\, d\omega. \tag{5.8.4}$$

Letting $\mathbf{b} \in L^1(\mathbb{R}; L^2(\Omega))$, $Y \in L^1(\mathbb{R}^+)$, consider the functional

$$\Phi_v(\mathbf{u}) = \tfrac{1}{2}\langle \nabla\mathbf{u}, (\mathbf{G}_0 + \mathbf{G}'\circ)\nabla\mathbf{u}\rangle_v - \langle \mathbf{u}, \mathbf{b}\rangle_v;$$

here we let

$$\mathbf{G}' \circ \nabla\mathbf{u} = \int_{-\infty}^{\infty} \mathbf{G}'(s)\nabla\mathbf{u}(t-s)\, ds$$

but take $\mathbf{G}'(s) = 0$ as $s < 0$. The functional Φ_v is well defined on $L^1(\mathbb{R}; H_0^1(\Omega))$. Now we show that Φ_v is stationary at the solution to the quasi-static problem expressed by (4.3.1)-(4.3.2).

THEOREM 5.8.3. *Let $\mathbf{b} \in L^1(\mathbb{R}; L^2(\Omega))$. Then $\mathbf{u}^0 \in L^1(\mathbb{R}; H_0^1(\Omega))$ is a stationary point of Φ_v for every $Y \in L^1(\mathbb{R}^+)$ if and only if $\mathbf{u}^0$ is a solution to the quasi-static problem.*

Proof. Let $\mathbf{u}^0 \in L^1(\mathbb{R}; H_0^1(\Omega))$ be a solution to the quasi-static problem with $\mathbf{b} \in L^1(\mathbb{R}; L^2(\Omega))$. Then, for any $\omega \in \mathbb{R}$, the Fourier transform $\mathbf{u}_F^0 \in H_0^1(\Omega)$ satisfies

$$\nabla \cdot \{[\mathbf{G}_0 + \mathbf{G}'_F(\omega)]\nabla\mathbf{u}_F^0(\omega)\} + \rho\mathbf{b}_F(\omega) = 0 \tag{5.8.5}$$

almost everywhere in Ω. Letting $L_\omega\mathbf{u} := \nabla \cdot \{[\mathbf{G}_0 + \mathbf{G}'_F(\omega)]\nabla\mathbf{u}\}$ by (5.8.4) and the divergence theorem we have

$$d\Phi_v(\mathbf{u}^0|\mathbf{v}) = -\int_{-\infty}^{\infty} \Upsilon(\omega)\, \Im(L_\omega\mathbf{u}_F^0(\omega) + \rho\mathbf{b}_F(\omega),\ \mathbf{v}_F(\omega))\, d\omega$$

for every $\mathbf{v} \in L^1(\mathbb{R}; H_0^1(\Omega))$ and $Y \in L^1(\mathbb{R}^+)$. Hence by (5.8.5) it follows that Φ_v is stationary at $\mathbf{u}^0$ in $L^1(\mathbb{R}; H_0^1(\Omega))$. Conversely, let Φ_v be stationary at $\mathbf{u}^0$ in $L^1(\mathbb{R}; H_0^1(\Omega))$. Then, on representing $\mathbf{u}$ as $\mathbf{u}^0 + h(t)\,\mathbf{w}(\mathbf{x})$, $h \in L^1(\mathbb{R})$, $\mathbf{w} \in H_0^1(\Omega)$, we have

$$0 = d\Phi_v(\mathbf{u}^0|h\,\mathbf{w}) = -2\int_0^\infty Y(\omega)[h_C(\omega)\,\Im(L_\omega \mathbf{u}^0(\omega) + \rho\mathbf{b}_F(\omega), \mathbf{w}) \\ - h_S(\omega)\,\Re(L_\omega \mathbf{u}^0(\omega) + \rho\mathbf{b}_F(\omega), \mathbf{w})]\,d\omega.$$

By the arbitrariness of $\mathbf{w}$, Yh_C, Yh_S, we obtain the real and imaginary part of (5.8.5) and then (5.8.5) as such. Accordingly, $\mathbf{u}^0$ is a solution to (5.8.5) for almost every value of ω. The uniqueness of the Fourier transform makes $\mathbf{u}^0$ the solution to the quasi-static problem. □

Recourse to the thermodynamic inequalities for $\mathbf{G}$ allows us to set up a saddle-point principle. In this regard, consider the space $W = L^2(\mathbb{R}; H_0^1(\Omega))$ and let the subspaces $W^+, W^- \subset W$ consist of even functions and odd functions, respectively. That is, any $\mathbf{u} \in W$ is the sum of

$$\mathbf{u}^+(t) = \tfrac{1}{2}[\mathbf{u}(t) + \mathbf{u}(-t)], \qquad \mathbf{u}^-(t) = \tfrac{1}{2}[\mathbf{u}(t) - \mathbf{u}(-t)].$$

Of course, W^+ and W^- are orthogonal to each other relative to the inner product of $L^2(\mathbb{R}; L^2(\Omega))$ and $W^+ \oplus W^- = W$. Further,

$$\mathbf{u}_F^+ = \mathbf{u}_C, \qquad \mathbf{u}_F^- = -i\mathbf{u}_S.$$

Then we can express the functional $\Phi_v(\mathbf{u})$ on $W^+ \times W^-$ as

$$\Psi_v(\mathbf{u}^+, \mathbf{u}^-) = \tfrac{1}{2}\langle\nabla(\mathbf{u}^+ + \mathbf{u}^-), (\mathbf{G}_0 + \mathbf{G}'\circ)\nabla(\mathbf{u}^+ + \mathbf{u}^-)\rangle_v - \langle(\mathbf{u}^+ + \mathbf{u}^-), \rho\mathbf{b}\rangle_v.$$

Here $\mathbf{b} \in L^2(\mathbb{R}; L^2(\Omega))$ and $Y \in C_0^\infty(\mathbb{R}^+)$. Owing to (5.8.3) and the bijectivity of the Fourier transform from $L^2(\Omega)$ into itself, the functional Ψ_v is well defined in $W^+ \times W^-$.

THEOREM 5.8.4. *Let $\mathbf{G}'$ satisfy* (3.4.1) *and* $\mathbf{b} \in L^2(\mathbb{R}; L^2(\Omega))$. *Then the pair $(\mathbf{u}^+, \mathbf{u}^-)$ is a saddle point of Ψ_v on $W^+ \times W^-$ for every $Y \in C_0^\infty(\mathbb{R}^+, \mathbb{R}^+)$ if and only if $\mathbf{u} = \mathbf{u}^+ + \mathbf{u}^-$ is a solution to the quasi-static problem. If, in addition,* (3.1.1) *and* (3.2.11) *hold, then the saddle point exists and is unique in W.*

Proof. If $\mathbf{u} \in W$ is a solution to the quasi-static problem with $\mathbf{b} \in L^2(\mathbb{R}; L^2(\Omega))$ then $\mathbf{u}_F \in W$ satisfies the complex-valued equation

$$\nabla \cdot \{[\mathbf{G}_0 + \mathbf{G}'(\omega)]\nabla\mathbf{u}_F(\omega)\} + \rho\mathbf{b}_F(\omega) = 0 \tag{5.8.6}$$

almost everywhere in $\Omega \times \mathbb{R}$. Hence $\mathbf{u}_C^+ \in W^+$ and $\mathbf{u}_S^- \in W^-$ satisfy almost everywhere the system of real-valued equations

$$\nabla \cdot \{[\mathbf{G}_0 + \mathbf{G}'_C(\omega)]\nabla\mathbf{u}_C(\omega) - \mathbf{G}'_S(\omega)\nabla\mathbf{u}_S(\omega)\} + \rho\mathbf{b}_C(\omega) = 0, \tag{5.8.7}$$

$$\text{(5.8.8)} \qquad \nabla \cdot \{[\mathbf{G}_0 + \mathbf{G}'_C(\omega)]\nabla \mathbf{u}_S(\omega) + \mathbf{G}'_S(\omega)\nabla \mathbf{u}_C(\omega)\} + \rho \mathbf{b}_S(\omega) = 0.$$

By (5.8.4) and some rearrangement we can write

$$\begin{aligned}\Psi_v(\mathbf{u}^+, \mathbf{u}^-) = \int_0^\infty Y(\omega)[(\nabla \mathbf{u}_S,\, \mathbf{G}'_S \nabla \mathbf{u}_S) &- (\nabla \mathbf{u}_C, \mathbf{G}'_S \nabla \mathbf{u}_C) \\ &- 2(\nabla \mathbf{u}_S, (\mathbf{G}_0 + \mathbf{G}'_C)\nabla \mathbf{u}_C) \\ &+ 2(\mathbf{u}_C, \rho \mathbf{b}_S) + 2(\mathbf{u}_S, \rho \mathbf{b}_C)](\omega)\, d\omega.\end{aligned}$$

Observe that for every $\mathbf{v}^+ \in W^+$ it is $\mathbf{v}_F^+ = v_C^+ \in W^+$ and for every $\mathbf{v}^- \in W^-$ it is $i\mathbf{v}_F^- = \mathbf{v}_S^- \in W^-$. Accordingly, we can write the differentials with respect to $\mathbf{u}^+$ and $\mathbf{u}^-$ as

$$d\Psi_v(\mathbf{u}^+ | \mathbf{v}^+, \mathbf{u}^-) = 2\int_0^\infty Y(\nabla \cdot \{[\mathbf{G}_0 + \mathbf{G}'_C]\nabla \mathbf{u}_S + \mathbf{G}'_S \nabla \mathbf{u}_C\} + \rho \mathbf{b}_S,\, \mathbf{v}_F^+)\, d\omega,$$

for every $\mathbf{v}^+ \in W^+$,

$$d\Psi_v(\mathbf{u}^+, \mathbf{u}^- | \mathbf{v}^-) = 2\int_0^\infty Y(\nabla \cdot \{[\mathbf{G}_0 + \mathbf{G}'_S]\nabla \mathbf{u}_C - \mathbf{G}'_S \nabla \mathbf{u}_S\} + \rho \mathbf{b}_C,\, i\mathbf{v}_F^-)\, d\omega,$$

for every $\mathbf{v}^- \in W^-$. Then by (5.8.7) and (5.8.8) we have

$$d\Psi_v(\mathbf{u}^+ | \mathbf{v}^+, \mathbf{u}^-) = 0, \qquad \forall \mathbf{v}^+ \in W^+,$$

$$d\Psi_v(\mathbf{u}^+, \mathbf{u}^- | \mathbf{v}^-) = 0, \qquad \forall \mathbf{v}^- \in W^-,$$

for every $Y \in C_0^\infty(\mathbb{R}^+, \mathbb{R}^+)$. Further, by (3.4.1) we have

$$d^2\Psi_v(\mathbf{u}^+ | \mathbf{v}^+, \mathbf{u}^-) = -2\int_0^\infty Y\left(\nabla \mathbf{v}_C^+, \mathbf{G}'_S \nabla \mathbf{v}_C^+\right) d\omega > 0, \qquad \forall \mathbf{v}^+ \in W^+,$$

$$d^2\Psi_v(\mathbf{u}^+, \mathbf{u}^- | \mathbf{v}^-) = 2\int_0^\infty Y\left(\nabla \mathbf{v}_S^-, \mathbf{G}'_S \nabla \mathbf{v}_S^-\right) d\omega < 0, \qquad \forall \mathbf{v}^- \in W^-,$$

for every $Y \in C_0^\infty(\mathbb{R}^+, \mathbb{R}^+)$. This proves that the pair $(\mathbf{u}^+, \mathbf{u}^-)$ is a saddle point on $W^+ \times W^-$ for every Ψ_v.

Conversely, assume that for every $Y \in C_0^\infty(\mathbb{R}^+, \mathbb{R}^+)$, $d\Psi_v(\mathbf{u}|\mathbf{v}) = 0$ and $d^2\Psi_v$ is positive or negative according to whether the increment is $\mathbf{v}^+$ or $\mathbf{v}^-$. By the arbitrariness of $\Upsilon \mathbf{v}_F^+$ in W^- and $i\Upsilon \mathbf{v}_F^-$ in W^+, it follows that $\mathbf{u}_F = \mathbf{u}_F^+ + \mathbf{u}_F^-$ satisfies (5.8.6) almost everywhere on $\Omega \times \mathbb{R}$. By the bijectivity of the Fourier transform from L^2 into L^2, $\mathbf{u}$ must be a solution to the quasi-static problem in W.

According to [47], if (3.1.1) and (3.2.11) hold then the quasi-static problem has one and only one solution. This in turn proves the last part of the theorem. $\square$

Based on the framework set up in this chapter, one might have expected that the thermodynamic inequalities ensure the extremum of the functional at the solution of the problem under consideration. We have shown that such is not the case and, as one can easily see, this feature is not strictly related to the quasi-static approximation. Rather, even in the dynamic case, a functional similar to Ψ_v has a saddle point at the solution.

5.9. Convexity. The relaxation functions that satisfy the dissipation principle make a correspondent functional convex. Since the dissipation principle is in fact a consequence of the conditions (3.1.1) and (3.4.1), the result of this section is that convexity is always associated with the thermodynamic restrictions (cf. [49]).

Let H be a convex set of a normed space, e.g., the fading memory space $\mathcal{H}$ of histories of the velocity $\dot{\mathbf{u}}$ and the velocity gradient $\nabla\dot{\mathbf{u}}$. Let $\dot{\boldsymbol{\xi}} = (\dot{\mathbf{u}}, \nabla\dot{\mathbf{u}})$ be any element of H. Indeed, it is the history of $\dot{\boldsymbol{\xi}}$ which is the element of H but, for formal simplicity, we write functionals on H in the form $\Psi(\dot{\boldsymbol{\xi}})$.

A functional Ψ is said convex if, for each pair of elements $\dot{\boldsymbol{\xi}}_1, \dot{\boldsymbol{\xi}}_2 \in H$, and each $\lambda \in (0,1)$ the inequality

$$\Psi(\lambda\dot{\boldsymbol{\xi}}_1 + (1-\lambda)\dot{\boldsymbol{\xi}}_2) \leq \lambda\Psi(\dot{\boldsymbol{\xi}}_1) + (1-\lambda)\Psi(\dot{\boldsymbol{\xi}}_2)$$

holds. As is well known, this inequality is equivalent to each of the conditions

$$\Psi(\dot{\boldsymbol{\xi}}_2) - \Psi(\dot{\boldsymbol{\xi}}_1) - d\Psi(\dot{\boldsymbol{\xi}}_1 | \dot{\boldsymbol{\xi}}_2 - \dot{\boldsymbol{\xi}}_1) \geq 0,$$

$$d^2\Psi(\dot{\boldsymbol{\xi}}_1 | \dot{\boldsymbol{\xi}}_2 - \dot{\boldsymbol{\xi}}_1) \quad \text{positive definite.}$$

By examining the second-order Gateaux derivative $d^2\Psi$ now, we show that the functional

$$\Psi(\dot{\boldsymbol{\xi}}) = \tfrac{1}{2}\int_{-\infty}^{t}\int_{-\infty}^{\tau} \nabla\dot{\mathbf{u}}(\tau)\cdot\mathbf{G}(\tau-\sigma)\nabla\dot{\mathbf{u}}(\sigma)\,d\sigma\,d\tau - \rho\int_{-\infty}^{t}\mathbf{b}(\tau)\cdot\dot{\mathbf{u}}(\tau)\,d\tau$$

is convex.

As it stands, the functional Ψ is scarcely useful because of the nonsymmetric role of the tensors $\nabla\dot{\mathbf{u}}$ in the first integral ($\tau \geq \sigma$). Here we recover the symmetry by extending the definition of $\mathbf{G}$ to $\mathbb{R}$ in the form

$$\mathbf{G}(\tau) = \mathbf{G}(|\,\tau\,|), \qquad \tau \in \mathbb{R}.$$

Then we can write

$$\Psi(\mathbf{u}) = \frac{1}{4}\int_{-\infty}^{t}\int_{-\infty}^{t} \nabla\dot{\mathbf{u}}(\tau)\cdot\mathbf{G}(|\,\tau-\sigma\,|)\nabla\dot{\mathbf{u}}(\sigma)\,d\sigma\,d\tau - \rho\int_{-\infty}^{t}\mathbf{b}(\tau)\cdot\dot{\mathbf{u}}(\tau)\,d\tau.$$

Hence, letting $\mathbf{u}+\nu\mathbf{w}\in H$, $\nu\in[0,\bar{\nu}], \bar{\nu}\in \mathbb{R}^+$, we have

$$d^2\Psi(\nabla\dot{\mathbf{u}}\mid\nabla\dot{\mathbf{w}})=\frac{1}{2}\int_{-\infty}^{t}\int_{-\infty}^{t}\nabla\dot{\mathbf{w}}(\tau)\cdot\mathbf{G}(\mid\tau-\sigma\mid)\nabla\dot{\mathbf{w}}(\sigma)\,d\sigma\,d\tau$$

where we have written $\nabla\dot{\mathbf{u}}$ and $\nabla\dot{\mathbf{w}}$ in the argument of $d^2\Psi$ to emphasize the independence of $\dot{\mathbf{u}}$ and $\dot{\mathbf{w}}$. Due to (3.1.3) we can write

$$d^2\Psi(\nabla\dot{\mathbf{u}}\mid\nabla\dot{\mathbf{w}})=\int_{-\infty}^{t}\nabla\dot{\mathbf{w}}(\tau)\cdot\boldsymbol{T}(\nabla\mathbf{w}^{\tau})\,d\tau.$$

This proves that the dissipation principle implies the convexity of the functional Ψ. By Theorem 3.4.3 this means also that, in H, the convexity of Ψ is a consequence of the thermodynamic condition (3.4.1) and the solid property (3.1.1).

Sometimes (cf., e.g., [89]) the functional $\Psi(\mathbf{u})$ is determined by appealing to the inverse problem of the calculus of variations. Nevertheless, in our minds there is no motivation for looking at Ψ, or rather

$$\psi_{\mathbf{E}}(\mathbf{E})=\frac{1}{2}\int_{-\infty}^{t}\int_{-\infty}^{\tau}\dot{\mathbf{E}}(\tau)\cdot\mathbf{G}(\tau-\sigma)\dot{\mathbf{E}}(\sigma)\,d\sigma\,d\tau,$$

as the potential for $\mathbf{T}$ in that neither $\mathbf{T}=\partial\psi_{\mathbf{E}}/\partial\mathbf{E}$ nor $\mathbf{T}=\partial\psi_{\mathbf{E}}/\partial\dot{\mathbf{E}}$.

5.10. Minimum principles for viscoelastic fluids. The minimum principles elaborated so far for viscoelastic solids have their nontrivial counterpart for viscoelastic fluids. To avoid being too repetitive, we are content with examining the analogue of the formulation, via a weight function, set up in §5.6. Both incompressible and compressible fluids are considered by following [65].

For formal convenience we write the constitutive equation (3.6.2) as

$$\mathbf{T}(t)=-p(t)\,\mathbf{1}+\int_{0}^{\infty}\mathbf{M}(s)\,\mathbf{L}(t-s)\,ds$$

where $\mathbf{M}:\mathbb{R}^+\to\mathrm{Lin(Sym)}$ and, in indicial notation,

$$M_{ijkl}=\mu(\delta_{ik}\,\delta_{jl}+\delta_{il}\,\delta_{jk})+\lambda\,\delta_{ij}\,\delta_{kl}.$$

In fact, if the fluid is regarded as incompressible, then λ has no effect at all.

We take the history of the velocity $\mathbf{v}$ to be known up to time $t=0$ for any $\mathbf{x}\in\Omega$ and let $\mathbf{v}$ be given at the boundary $\partial\Omega$ for any time $t\geq 0$. If $\mathbf{v}$ is known in $\Omega\times\mathbb{R}^-$ then we can write

$$\mathbf{T}(t)=-p\,\mathbf{1}+\int_{0}^{t}\mathbf{M}(s)\,\mathbf{L}(t-s)\,ds+\mathbf{T}_0(t)$$

where

$$\mathbf{T}_0(t) = \int_t^{\infty} \mathbf{M}(s)\, \mathbf{L}(t-s)\, ds$$

is known. The boundary $\partial\Omega$ of Ω is taken to consist of two disjoint subsets $\partial\Omega_v$, $\partial\Omega_T$, namely

$$\partial\Omega_v \cup \partial\Omega_T = \partial\Omega, \qquad \partial\Omega_v \cap \partial\Omega_T = \emptyset,$$

with boundary data for $\mathbf{v}$ on $\partial\Omega_v$ and for $\mathbf{T}$ on $\partial\Omega_T$.

Consistent with the linear approximation inherent in the model of viscoelasticity, we let $\dot{\mathbf{v}} \simeq \partial\mathbf{v}(\mathbf{x},t)/\partial t$. As usual we omit writing the dependence of the fields on the position $\mathbf{x}$. Then we consider the dynamics of the viscoelastic incompressible fluid subject to initial-boundary data through the following problem $\mathrm{P_I}$:[33]

$$(5.10.1) \qquad \rho\dot{\mathbf{v}} = -\nabla p + \int_0^t \mu(s)\Delta\mathbf{v}(t-s)\, ds + \rho\mathbf{f},$$

$$(5.10.2) \qquad \nabla\cdot\mathbf{v} = 0 \quad \text{on } \Omega\times\mathbb{R}^+,$$

$$(5.10.3) \qquad \mathbf{v} = 0 \quad \text{on } \partial\Omega_v \times \mathbb{R}^+, \qquad \mathbf{T}\mathbf{n} = \hat{\mathbf{t}} \quad \text{on } \partial\Omega_T\times\mathbb{R}^+,$$

$$(5.10.4) \qquad \mathbf{v}(0) = \mathbf{v}_0 \quad \text{on } \bar{\Omega},$$

where $\rho\mathbf{f} = \rho\mathbf{b} + \nabla\cdot\mathbf{T}_0$ and $\mathbf{b}$, as usual, is the body force. Any solution to the problem $\mathrm{P_I}$ is a pair of vector and scalar fields $\{\mathbf{v}, p\}$ on $\Omega\times\mathbb{R}^+$ satisfying (5.10.1)–(5.10.4). Of course, the initial velocity field $\mathbf{v}_0$ is required to satisfy the incompressibility constraint (5.10.2). Moreover, letting

$$J_v^1(\Omega) = \{\mathbf{v}\in H^1(\Omega) : \ \nabla\cdot\mathbf{v} = 0 \text{ on } \Omega, \ \mathbf{v} = 0 \text{ on } \partial\Omega_v\},$$

we require that $\mathbf{v}(t) \in J_v^1(\Omega)$ almost everywhere on $\mathbb{R}^+$, which makes (5.10.2) and $(5.10.3)_1$ identically true.

For later convenience, we recall that the shear relaxation function μ is required to satisfy the thermodynamic inequality (3.7.5), namely the half-range Fourier cosine transform $\mu_c(\omega)$ is strictly positive for any $\omega\in\mathbb{R}^{++}$. Through the following lemma, the inequality (3.7.5) gives a corresponding inequality for the Laplace transform μ_L.

---

[33]Here we generalize the investigation performed in [18] which corresponds to the case $\partial\Omega_T = \emptyset$.

LEMMA 5.10.1. *Let $\mu \in L^1(\mathbb{R}^+)$ and $p \in \mathbb{C}^{++}$. The Laplace transform μ_L and the half-range cosine transform μ_c are related by*

$$\mu_L(p) = \frac{2}{\pi}\int_0^\infty \frac{p\mu_c(\omega)}{p^2+\omega^2}\,d\omega, \qquad p \in \mathbb{C}^{++}.$$

Proof. By Lemma 4.8.1, if $\phi, \psi \in L^1(\mathbb{R})$ and $\phi\psi \in L^1(\mathbb{R})$ then

$$\int_{-\infty}^{\infty} \phi(u)\,\psi(u)\exp(-i\beta u)\,du = \frac{1}{2\pi}\int_{-\infty}^{\infty} \phi_F(\omega)\,\psi_F(\beta-\omega)\,d\omega,$$

namely (4.8.11), provided that the right-hand side is continuous with respect to $\beta \in \mathbb{R}$. Letting $\alpha \in \mathbb{R}^{++}$, make the identifications

$$\phi(u) = \mu(|u|), \qquad u \in \mathbb{R}, \qquad \psi(u) = \begin{cases} 0, & u < 0, \\ \exp(-\alpha u), & u \geq 0. \end{cases}$$

Both ϕ and ψ belong to $L^1(\mathbb{R})$ and

$$\phi_F(\omega) = 2\,\mu_c(\omega), \qquad \psi_F(\omega) = \frac{1}{\alpha+i\omega}.$$

Hence

$$\phi_F(\omega)\,\psi_F(\beta-\omega) = \frac{2\,\mu_c(\omega)}{\alpha+i(\beta-\omega)}.$$

The right-hand side is absolutely integrable in ω and continuous with respect to β. Then (4.8.11) is applicable. It follows that

$$\int_0^\infty \mu(u)\exp[-(\alpha+i\beta)u]du = \frac{1}{\pi}\int_{-\infty}^{\infty} \frac{\mu_c(\omega)}{\alpha+i(\beta-\omega)}d\omega.$$

Then we can write

$$\mu_L(\alpha+i\beta) = \frac{1}{\pi}\int_{-\infty}^{\infty} \frac{[(\alpha+i(\beta+\omega)]\mu_c(\omega)}{(\alpha+i\beta)^2+\omega^2}d\omega, \qquad \alpha \in \mathbb{R}^{++}.$$

Since $\mu_c(\omega)$ is even in ω the desired result follows. □

As an immediate consequence, letting $\beta = 0$ we obtain from (3.7.5) that

(5.10.5) $$\mu_L(\alpha) > 0, \qquad \alpha \in \mathbb{R}^{++}.$$

By analogy with §5.6 we let

$$y(t) := \int_0^\infty Y(\alpha)\exp(-\alpha t)\,d\alpha, \qquad Y \in C_0^\infty(\mathbb{R}^+)$$

and make use of the bilinear form

$$(v,w)_y = \int_0^\infty \int_0^\infty y(t+\tau) \int_\Omega v(t)w(\tau)\,dx\,d\tau\,dt$$

for any pair of functions $v, w \in L^2_L(\mathbb{R}^+; L^2(\Omega))$. Then we consider the functional

$$\begin{aligned}\Upsilon_y(\mathbf{v}) =& \tfrac{1}{2}[\rho(\dot{\mathbf{v}},\mathbf{v})_y + (\mathbf{M} * \nabla\mathbf{v}, \nabla\mathbf{v})_y] - \rho(\mathbf{f},\mathbf{v})_y \\ &+ \rho\int_0^\infty y(t)\int_\Omega [\tfrac{1}{2}\mathbf{v}(0) - \mathbf{v}_0]\cdot\mathbf{v}(t)\,dx\,dt \\ &- \int_0^\infty\int_0^\infty y(t+\tau)\int_{\partial\Omega_T} \mathbf{v}(\tau)\cdot\hat{\mathbf{t}}_0(t)\,d\tau\,dt\,da\end{aligned}$$

where $\hat{\mathbf{t}}_0 = \hat{\mathbf{t}} - \mathbf{T}_0\mathbf{n}$. It is assumed that $\mathbf{f} \in L^2_L(\mathbb{R}^+; L^2(\Omega))$ and $\hat{\mathbf{t}}_0 \in L^2_L(\mathbb{R}^+; L^2(\partial\Omega_T))$.

In order to prove the desired minimum property for the functional Υ_y, we make use of the following lemmas.

LEMMA 5.10.2. *If the functions $u, \dot{u}, w, \dot{w}$ on $\mathbb{R}^+$ are Laplace transformable on $\mathbb{R}^+$, then*

$$\begin{aligned}\int_0^\infty\int_0^\infty y(t+\tau)\dot{u}(t)w(\tau)\,dt\,d\tau =& \int_0^\infty\int_0^\infty y(t+\tau)u(t)\dot{w}(\tau)\,dt\,d\tau \\ &- \int_0^\infty y(t)u(0)w(t)\,dt + \int_0^\infty y(t)u(t)w(0)\,dt.\end{aligned} \tag{5.10.6}$$

Proof. By the definition of y we have

$$\int_0^\infty\int_0^\infty y(t+\tau)v(t)w(\tau)\,d\tau\,dt = \int_0^\infty Y(\alpha)v_L(\alpha)w_L(\alpha)\,d\alpha.$$

Now, if $v = \dot{u}$ then $v_L(\alpha) = \alpha u_L(\alpha) - u(0)$. Accordingly,

$$\dot{u}_L(\alpha)w_L(\alpha) = -u(0)w_L(\alpha) + u_L(\alpha)[w(0) + \dot{w}_L(\alpha)]$$

and hence

$$\begin{aligned}\int_0^\infty Y(\alpha)\dot{u}_L(\alpha)w_L(\alpha)\,d\alpha =& -\int_0^\infty y(t)u(0)w(t)\,dt + \int_0^\infty y(t)u(t)w(0)\,dt \\ &+ \int_0^\infty\int_0^\infty y(t+\tau)u(t)\dot{w}(\tau)\,d\tau\,dt.\end{aligned}$$

The relation (5.10.6) then follows. □

LEMMA 5.10.3. *If* $\mathbf{h} \in C^1(\Omega \times \mathbb{R}^+)$ *is such that* $\nabla \cdot \mathbf{h} = 0$ *on* $\Omega \times \mathbb{R}^+$ *and* $\mathbf{h} \cdot \mathbf{n} = 0$ *on* $\partial\Omega_v \times \mathbb{R}^+$, *then, for any scalar field* $q \in C^1(\Omega \times \mathbb{R}^+)$,

$$\int_\Omega \nabla q(t) \cdot \mathbf{h}(\tau)\, dx = \int_{\partial\Omega_T} q(t)\, \mathbf{h}(\tau) \cdot \mathbf{n}\, da, \qquad t, \tau \in \mathbb{R}^+.$$

Proof. Since $\nabla \cdot \mathbf{h} = 0$ then $\nabla q(t) \cdot \mathbf{h}(\tau) = \nabla \cdot [q(t)\, \mathbf{h}(\tau)]$. The divergence theorem and the vanishing of $\mathbf{h} \cdot \mathbf{n}$ on $\partial\Omega_v$ provide the result. □

Now we can prove two theorems about the minimum of the functional Υ_y; of course it is understood that μ meets the thermodynamic inequality (3.7.5) and then, by Lemma 5.10.1, μ_L meets (5.10.5). To this end it is worth considering the space

$$\mathcal{Q}_v = L^2_L(\mathbb{R}^+; J^1_v(\Omega)) \cap H^1_L(\mathbb{R}^+; L^2(\Omega)).$$

THEOREM 5.10.1. *If* $\mathbf{v}^0 \in \mathcal{Q}_v$ *is a solution to the problem* P_1, *then for every function* $Y \in C_0^\infty(\mathbb{R}^+, \mathbb{R}^+)$ *the functional* $\Upsilon_y(\mathbf{v})$ *has a minimum at* $\mathbf{v}^0$ *on* $\mathcal{Q}_v$.

Proof. Let $\mathbf{v}^0 \in \mathcal{Q}_v$ and represent any element $\mathbf{v} \in \mathcal{Q}_v$ as $\mathbf{v}^0 + \mathbf{h}$, where $\mathbf{h} \in \mathcal{Q}_v$. Then we can write

$$\begin{aligned}
\Upsilon_y(\mathbf{v}^0{+}\mathbf{h}) - \Upsilon_y(\mathbf{v}^0) &= \tfrac{1}{2}[\rho(\dot{\mathbf{v}}^0, \mathbf{h})_y + \rho(\dot{\mathbf{h}}, \mathbf{v}^0)_y + \rho(\dot{\mathbf{h}}, \mathbf{h})_y - 2\,\rho(\mathbf{f}, \mathbf{h})_y \\
&\quad + (\mathbf{M} * \nabla\mathbf{h}, \nabla\mathbf{v}^0)_y + (\mathbf{M} * \nabla\mathbf{v}^0, \nabla\mathbf{h})_y + (\mathbf{M} * \nabla\mathbf{h}, \nabla\mathbf{h})_y] \\
&\quad + \rho \int_0^\infty y(t) \int_\Omega \{[\tfrac{1}{2}\mathbf{v}^0(0) - \mathbf{v}_0] \cdot \mathbf{h}(t) + \tfrac{1}{2}\mathbf{h}(0) \cdot [\mathbf{v}^0(t) + \mathbf{h}(t)]\} dx\, dt \\
&\quad - \int_0^\infty \int_0^\infty y(t+\tau) \int_{\partial\Omega_T} \hat{\mathbf{t}}_0(t) \cdot \mathbf{h}(\tau)\, da\, d\tau\, dt.
\end{aligned}$$

By virtue of Lemma 5.10.2 and the property (5.6.2) we have

$$\begin{aligned}
\Upsilon_y(\mathbf{v}^0 + \mathbf{h}) - \Upsilon_y(\mathbf{v}^0) &= \rho(\dot{\mathbf{v}}^0, \mathbf{h})_y + (\mathbf{M} * \nabla\mathbf{v}^0, \nabla\mathbf{h})_y - \rho(\mathbf{f}, \mathbf{h})_y \\
&\quad + \rho \int_0^\infty y(t) \int_\Omega \{[\mathbf{v}^0(0) - \mathbf{v}_0] \cdot \mathbf{h}(t) + \tfrac{1}{2}\mathbf{h}(0) \cdot \mathbf{h}(t)\} dx\, dt \\
&\quad - \int_0^\infty \int_0^\infty y(t+\tau) \int_{\partial\Omega_T} \hat{\mathbf{t}}_0(t) \cdot \mathbf{h}(\tau)\, da\, d\tau\, dt \\
&\quad + \tfrac{1}{2}[\rho(\dot{\mathbf{h}}, \mathbf{h})_y + (\mathbf{M} * \nabla\mathbf{h}, \nabla\mathbf{h})_y].
\end{aligned}$$

Observe that, because $\mathbf{h} = 0$ on $\partial\Omega_v \times \mathbb{R}^+$,

$$\begin{aligned}
\int_\Omega (\mathbf{M} * \nabla\mathbf{v}^0)(t) \cdot \nabla\mathbf{h}(\tau) dx &= \int_{\partial\Omega_T} \mathbf{n} \cdot (\mathbf{M} * \nabla\mathbf{v}^0)(t)\, \mathbf{h}(\tau) da \\
&\quad - \int_\Omega (\mu * \Delta\mathbf{v}^0)(t) \cdot \mathbf{h}(\tau) dx.
\end{aligned}$$

Because $\mathbf{v}^0$ is the solution to the problem (5.10.1)-(5.10.4), the vanishing of $\mathbf{v}^0(0) - \mathbf{v}_0$, the use of Lemma 5.10.3 and account of the boundary conditions (5.10.3) yield

$$\Upsilon_y(\mathbf{v}^0+\mathbf{h})-\Upsilon(\mathbf{v}^0) = \tfrac{1}{2}\Big[\rho(\dot{\mathbf{h}}, \mathbf{h})_y+(\mathbf{M}*\nabla\mathbf{h}, \nabla\mathbf{h})_y+\rho\int_0^\infty y(t)\int_\Omega \mathbf{h}(0)\cdot\mathbf{h}(t)\, dx\, dt\Big].$$

Now, $(\mathbf{M} * \nabla\mathbf{h}, \nabla\mathbf{h})_y = 2(\mu * \mathrm{sym}\nabla\mathbf{h},\ \mathrm{sym}\nabla\mathbf{h})_y$. Hence, by the choice of the weight function y, we obtain

$$\Upsilon_y(\mathbf{v}^0+\mathbf{h})-\Upsilon_y(\mathbf{v}^0) = \tfrac{1}{2}\int_0^\infty Y(\alpha)\int_\Omega [\rho\,\alpha\,|\mathbf{h}_L(\alpha)|^2+2\,\mu_L(\alpha)\,|\mathrm{sym}\nabla\mathbf{h}_L(\alpha)|^2]dx\, d\alpha.$$

In view of (5.10.5) we conclude that $\Upsilon_y(\mathbf{v}^0 + \mathbf{h}) - \Upsilon_y(\mathbf{v}^0)$ vanishes only at $\mathbf{h} = 0$, that is, at $\mathbf{v} = \mathbf{v}^0$. □

Really, since Y may have a compact support, we can conclude that $\mathbf{h}_L = 0$ on the pertinent subinterval. However, real Laplace transforms can be considered as restrictions to the real axis of analytic functions on the complex plane. The fact that analytic functions vanishing on an interval of the real axis vanish on $\mathbb{R}^+$ too provides the vanishing of $\mathbf{h}_L$ on $\Omega \times \mathbb{R}^+$ and then that of $\mathbf{h}$ on $\Omega \times \mathbb{R}^+$.

THEOREM 5.10.2. *If* $\mathbf{v}^0 \in \mathcal{Q}_v$ *and*

$$\Upsilon_y(\mathbf{v}^0) \le \Upsilon_y(\mathbf{v}), \qquad \mathbf{v} \in \mathcal{Q}_v,$$

for every $Y \in C_0^\infty(\mathbb{R}^+, \mathbb{R}^+)$, *then* $\mathbf{v}^0$ *is the unique solution to the problem* P_1.

Proof. Any $\mathbf{v} \in \mathcal{Q}_v$ can be expressed as $\mathbf{v} = \mathbf{v}^0 + \eta\mathbf{h}$ with $\mathbf{h} \in \mathcal{Q}_v$, $\eta \in \mathbb{R}$. By direct calculation we have

$$\begin{aligned}
&\Upsilon_y(\mathbf{v}^0 + \eta\mathbf{h}) - \Upsilon_y(\mathbf{v}^0) = \\
&\eta\Big\{\tfrac{1}{2}[\rho(\dot{\mathbf{v}}^0, \mathbf{h})_y + \rho(\dot{\mathbf{h}}, \mathbf{v}^0)_y + (\mathbf{M} * \nabla\mathbf{v}^0, \nabla\mathbf{h})_y + (\mathbf{M} * \nabla\mathbf{h}, \nabla\mathbf{v}^0)_y] - \rho(\mathbf{f}, \mathbf{h})_y \\
&\qquad + \rho\int_0^\infty y(t)\int_\Omega \{[\tfrac{1}{2}\mathbf{v}^0(0) - \mathbf{v}_0]\cdot\mathbf{h}(t) + \tfrac{1}{2}\mathbf{h}(0)\cdot\mathbf{v}^0(t)\}dx\, dt \\
&\qquad - \int_0^\infty\int_0^\infty y(t+\tau)\int_{\partial\Omega_T} \hat{\mathbf{t}}_0(t)\cdot\mathbf{h}(\tau)\, da\, d\tau\, dt\Big\} \\
&+ \tfrac{1}{2}\eta^2\Big\{\rho(\dot{\mathbf{h}}, \mathbf{h})_y + (\mathbf{M} * \nabla\mathbf{h}, \nabla\mathbf{h})_y + \rho\int_0^\infty y(t)\int_\Omega \mathbf{h}(0)\cdot\mathbf{h}(t)\, dx\, dt\Big\}.
\end{aligned}$$

Accordingly, Υ_y is minimal at $\mathbf{v}^0$ only if the coefficient of η vanishes while that of η^2 is nonnegative. Then by means of Lemma 5.10.2 and integrations by parts we obtain

$$\begin{aligned}
&(\rho\dot{\mathbf{v}}^0 - \mu * \Delta\mathbf{v}^0 - \rho\mathbf{f},\, \mathbf{h})_y + \rho\int_0^\infty y(t)\int_\Omega [\mathbf{v}^0(0) - \mathbf{v}(0)]\cdot\mathbf{h}(t)\, dx\, dt \\
&+ \int_0^\infty\int_0^\infty y(t+\tau)\int_{\partial\Omega_T} [-\hat{\mathbf{t}}_0(t) + \mathbf{n}(\mathbf{M} * \nabla\mathbf{v}^0)(t)]\cdot\mathbf{h}(\tau)\, da\, d\tau\, dt = 0,
\end{aligned} \tag{5.10.7}$$

$$(5.10.8)\qquad (\rho\dot{\mathbf{h}}, \mathbf{h})_y + (\mathbf{M} * \nabla\mathbf{h}, \nabla\mathbf{h})_y + \rho\int_0^\infty y(t)\int_\Omega \mathbf{h}(0)\cdot\mathbf{h}(t)dx\,dt \geq 0.$$

The left-hand side of (5.10.8) can be written as

$$\int_0^\infty Y(\alpha)\int_\Omega [\rho\,\alpha\,|\mathbf{h}_L(\alpha)|^2 + 2\,\mu_L(\alpha)\,|\mathrm{sym}\nabla\mathbf{h}_L(\alpha)|^2]\,d\alpha\,dx$$

and shows that, by (5.10.5), Υ_y has a strict minimum at $\mathbf{v}^0$, namely the minimum is unique. In view of Lemma 5.10.3, the divergence-free condition on $\mathbf{h}$ and the vanishing of $\mathbf{h}$ on $\partial\Omega_v$ allows us to add the (zero) quantity

$$\int_\Omega \nabla\Pi(t)\cdot\mathbf{h}(\tau)dx - \int_{\partial\Omega_T}\Pi(t)\mathbf{h}(\tau)\cdot\mathbf{n}da$$

on the left-hand side of (5.10.7). Hence (5.10.7) yields

$$\rho\,\alpha\,\mathbf{v}_L^0 + \nabla\Pi_L - \mu_L\,\Delta\mathbf{v}_L^0 - \rho\mathbf{f}_L - \rho\mathbf{v}_0 = 0 \quad \text{on } \Omega$$

along with

$$(\hat{\mathbf{t}}_0)_L = -\Pi_L\,\mathbf{n} + \mathbf{n}(\mathbf{M}_L\nabla\mathbf{v}_L^0) \quad \text{on } \partial\Omega_T.$$

So $\mathbf{v}_L^0$ is the unique solution to the Laplace transformed problem and $\mathbf{v}^0$ is the unique solution to the problem (5.10.1)-(5.10.4). □

Now let the fluid be compressible. We take the balance equations to be given by their linear approximations and the history of $\mathbf{v}$ and ρ to be known up to time $t = 0$ for any $\mathbf{x} \in \Omega$. Accordingly, we let ρ_0 be the average mass density and set $\varphi = \rho - \rho_0$. Then, letting again $\dot{} = \partial/\partial t$, we write

$$(5.10.9)\qquad \rho_0\dot{\mathbf{v}} = -\Pi_\rho\nabla\varphi + \nabla\cdot\int_0^t \mathbf{M}(s)\nabla\mathbf{v}(t-s)\,ds + \rho_0\tilde{\mathbf{f}},$$

$$(5.10.10)\qquad \dot{\varphi} + \rho_0\nabla\cdot\mathbf{v} = 0,$$

where

$$\rho_0\tilde{\mathbf{f}} = \rho_0\mathbf{b} + \nabla\cdot\int_t^\infty \mathbf{M}(s)\nabla\mathbf{v}(t-s)\,ds$$

and Π_ρ stands for the reference value $\Pi_\rho(\rho_0)$. The boundary data (5.10.3) hold along with the initial data

$$(5.10.11)\qquad \mathbf{v}(0) = \mathbf{v}_0, \qquad \varphi(0) = \varphi_0 \qquad \text{on } \bar{\Omega}.$$

Integration of (5.10.10) over Ω and the divergence theorem yield

$$\int_\Omega \dot{\varphi}(t)\,dx = -\rho_0\int_{\partial\Omega}\mathbf{v}(t)\cdot\mathbf{n}\,da$$

whence, by $(5.10.3)_1$,

$$\int_\Omega \dot{\varphi}(t)\,dx = -\rho_0 \int_{\partial\Omega_T} \mathbf{v}(t)\cdot\mathbf{n}\,da.$$

Then mass conservation in the form

$$\int_\Omega \rho(t)\,dx = \int_\Omega \rho_0\,dx, \qquad t\in\mathbb{R}^+,$$

yields the constraints

$$(5.10.12)\quad \int_\Omega \varphi(t)\,dx = 0, \qquad t\in\mathbb{R}^+, \qquad \int_{\partial\Omega_T} \mathbf{v}(t)\cdot\mathbf{n}\,da = 0, \qquad t\in\mathbb{R}^+.$$

In particular, we can say that φ_0 has a vanishing average value in Ω and $\mathbf{v}_0$ has a vanishing flux across $\partial\Omega_T$. On the whole (5.10.9)–(5.10.12) and (5.10.3) constitute the dynamic problem $\mathrm{P_f}$.

It is a straightforward matter to verify that the functional

$$\begin{aligned}\Xi_y(\mathbf{v},\varphi) &= \tfrac{1}{2}\Big[\rho_0(\dot{\mathbf{v}},\mathbf{v})_y - \frac{\Pi_\rho}{\rho_0}(\dot{\varphi},\varphi)_y + (\mathbf{M}*\nabla\mathbf{v},\nabla\mathbf{v})_y\Big]\\ &\quad + \Pi_\rho(\nabla\varphi,\mathbf{v})_y - \rho_0(\tilde{\mathbf{f}},\mathbf{v})_y\\ &\quad - \int_0^\infty\int_0^\infty y(t+\tau)\int_{\partial\Omega_T}\mathbf{v}(t)\cdot\hat{\mathbf{t}}_0(\tau)\,da\,d\tau\,dt\\ &\quad + \int_0^\infty y(t)\int_\Omega\Big[\rho_0(\tfrac{1}{2}\mathbf{v}(0)-\mathbf{v}_0)\cdot\mathbf{v}(t) - \frac{\Pi_\rho}{\rho_0}(\tfrac{1}{2}\varphi(0)-\varphi_0)\varphi(t)\Big]\,dx\,dt\end{aligned}$$

is stationary at the solution $\mathbf{v},\varphi$ to the problem $\mathrm{P_f}$. Yet, owing to the term $(\nabla\varphi,\mathbf{v})_y$ and the $-$ sign in front of $(\dot{\varphi},\varphi)_y$, we cannot find a definite form for the possible extremum. To remove the difficulty, we consider, in a sense, a functional for one equation only but account for the other one as a constraint.

By (5.10.10) we have

$$\varphi(t) = \varphi_0 - \rho_0\int_0^t (\nabla\cdot\mathbf{v})(s)\,ds.$$

For formal convenience we write

$$\varphi = \varphi_0 - \rho_0 * \nabla\cdot\mathbf{v}.$$

Then, letting

$$N_{ijkl} = M_{ijkl} + \Pi_\rho\rho_0\delta_{ij}\delta_{kl}$$

and

$$\rho_0 \mathbf{f} = \rho_0 \tilde{\mathbf{f}} - \Pi_\rho \nabla \varphi_0,$$

we can express the problem $\mathrm{P_f}$ in the form, $\mathrm{P_v}$ say,

$$\rho_0 \dot{\mathbf{v}} = \nabla \cdot (\mathbf{N} * \nabla \mathbf{v}) + \rho_0 \mathbf{f},$$

$$\mathbf{v}(0) = \mathbf{v}_0 \quad \text{in } \bar{\Omega}, \qquad \mathbf{v} = 0 \quad \text{on } \partial\Omega_v \times \mathbb{R}^+,$$

$$\mathbf{T}\,\mathbf{n} = \hat{\mathbf{t}} \quad \text{on } \partial\Omega_T \times \mathbb{R}^+, \qquad \int_{\partial\Omega_T} \mathbf{v} \cdot \mathbf{n}\, da = 0 \quad \text{on } \mathbb{R}^+.$$

In the form $\mathrm{P_v}$, involving only the velocity field $\mathbf{v}$, we can determine an extremum principle. In the space

$$H^1_{\mathbf{v}}(\Omega) = \left\{ \mathbf{v} \in H^1(\Omega) : \ \mathbf{v} = 0 \quad \text{on} \quad \partial\Omega_v, \quad \int_{\Omega_T} \mathbf{v} \cdot \mathbf{n}\, da = 0 \right\},$$

consider the functional

$$\begin{aligned} \Theta_y(\mathbf{v}) = {} & \tfrac{1}{2}[\rho_0(\dot{\mathbf{v}}, \mathbf{v})_y + (\mathbf{N} * \nabla \mathbf{v}, \nabla \mathbf{v})_y] - \rho_0(\mathbf{f}, \mathbf{v})_y \\ & + \rho_0 \int_0^\infty y(t) \int_\Omega [\tfrac{1}{2}\mathbf{v}(0) - \mathbf{v}_0] \cdot \mathbf{v}(t)\, dx\, dt \\ & - \int_0^\infty \int_0^\infty y(t+\tau) \int_{\partial\Omega_T} \mathbf{v}(\tau) \cdot \hat{\mathbf{t}}_0(t)\, da\, d\tau\, dt \end{aligned}$$

where $\hat{\mathbf{t}}_0 = \hat{\mathbf{t}} - \mathbf{T}_0 \mathbf{n}$ and

$$\mathbf{T}_0(t) = -[\Pi(\rho_0) + \Pi_\rho \varphi_0]\mathbf{1} + \int_t^\infty \mathbf{M}(s)\mathbf{L}(t-s)\, ds.$$

Following a customary, usually tacit, assumption, we let $\Pi_\rho > 0$. Moreover, by the thermodynamic inequalities (3.7.3) and the analogue of Lemma 5.10.1 we have

$$(5.10.13) \qquad \mu_L(\alpha) > 0, \qquad \kappa_L(\alpha) > 0, \qquad \alpha \in \mathbb{R}^+.$$

In a moment we need the following result.

LEMMA 5.10.4. *If u, v on $\mathbb{R}^+$ are Laplace transformable on $\mathbb{R}^+$ and $U(t) = \int_0^t u(\tau)\, d\tau$, $V(t) = \int_0^t v(\tau)\, d\tau$, then*

$$(U, v)_y = (u, V)_y.$$

Proof. Since $U_L(\alpha) = u_L(\alpha)/\alpha$, by the definition of y we have

$$\int_0^\infty \int_0^\infty y(t+\tau)\, U(t)\, v(\tau)\, d\tau\, dt = \int_0^\infty Y(\alpha)\, \frac{1}{\alpha} u_L(\alpha)\, v_L(\alpha)\, d\alpha.$$

The observation that $v_L(\alpha)/\alpha = V_L(\alpha)$ provides the desired result. □

Now we can show a minimum property of the functional Θ_y.

THEOREM 5.10.3. *If* $\mathbf{v}^0 \in \bar{\mathcal{Q}}_v = L^2_L(\mathbb{R}^+; H^1_v(\Omega)) \cap H^1_L(\mathbb{R}^+; L^2(\Omega))$, *and* $\mathbf{v}^0$ *meets the problem* $\mathrm{P_v}$, *then, for every function* $Y \in C_0^\infty(\mathbb{R}^+, \mathbb{R}^+)$, *the functional* $\Theta_y(\mathbf{v})$ *has a minimum at* $\mathbf{v}^0$ *on* $\bar{\mathcal{Q}}_v$.

Proof. Represent any element $\mathbf{v} \in \bar{\mathcal{Q}}_v$ as $\mathbf{v}^0 + \mathbf{h}$, $\mathbf{h} \in \bar{\mathcal{Q}}_v$. Then we can write

$$\begin{aligned}\Theta_y(\mathbf{v}^0+\mathbf{h}) - \Theta_y(\mathbf{v}^0) &= \tfrac{1}{2}[\rho_0(\dot{\mathbf{v}}^0, \mathbf{h})_y + \rho_0(\dot{\mathbf{h}}, \mathbf{v}^0)_y + \rho_0(\dot{\mathbf{h}}, \mathbf{h})_y - 2\rho_0(\mathbf{f}, \mathbf{h})_y \\ &\quad + (\mathbf{N} * \nabla\mathbf{h}, \nabla\mathbf{v}^0)_y + (\mathbf{N} * \nabla\mathbf{v}^0, \nabla\mathbf{h})_y + (\mathbf{N} * \nabla\mathbf{h}, \nabla\mathbf{h})_y] \\ &\quad + \rho_0 \int_0^\infty y(t) \int_\Omega [(\tfrac{1}{2}\mathbf{v}^0(0) - \mathbf{v}_0) \cdot \mathbf{h}(t) + \tfrac{1}{2}\mathbf{h}(0) \cdot (\mathbf{v}^0(t) + \mathbf{h}(t))] dx\, dt \\ &\quad - \int_0^\infty \int_0^\infty y(t+\tau) \int_{\partial\Omega_T} \hat{\mathbf{t}}_0(t) \cdot \mathbf{h}(\tau)\, da\, d\tau\, dt.\end{aligned}$$

By virtue of (5.10.6), (5.6.2), and Lemma 5.10.4 we have

$$\begin{aligned}\Theta_y(\mathbf{v}^0 + \mathbf{h}) - \Theta_y(\mathbf{v}^0) &= \rho_0(\dot{\mathbf{v}}^0, \mathbf{h})_y + (\mathbf{N} * \nabla\mathbf{v}^0, \nabla\mathbf{h})_y - \rho_0(\mathbf{f}, \mathbf{h})_y \\ &\quad + \rho_0 \int_0^\infty y(t) \int_\Omega \{[\mathbf{v}^0(0) - \mathbf{v}_0] \cdot \mathbf{h}(t) + \tfrac{1}{2}\mathbf{h}(0) \cdot \mathbf{h}(t)\} dx\, dt \\ &\quad - \int_0^\infty \int_0^\infty y(t+\tau) \int_{\partial\Omega_T} \hat{\mathbf{t}}_0(t) \cdot \mathbf{h}(\tau)\, da\, d\tau\, dt \\ &\quad + \tfrac{1}{2}[\rho_0(\dot{\mathbf{h}}, \mathbf{h})_y + (\mathbf{N} * \nabla\mathbf{h}, \nabla\mathbf{h})_y].\end{aligned}$$

Integration by parts, the divergence theorem, and some rearrangement yield

$$\begin{aligned}(\mathbf{N} * \nabla\mathbf{v}^0, \nabla\mathbf{h})_y &= -(\nabla \cdot \mathbf{N} * \nabla\mathbf{v}^0, \mathbf{h})_y \\ &\quad + \int_0^\infty \int_0^\infty y(t+\tau) \int_{\partial\Omega_T} [\mathbf{n}\,\mathbf{N} * \nabla\mathbf{v}^0](t) \cdot \mathbf{h}(\tau)\, da\, d\tau\, dt.\end{aligned}$$

Because $\mathbf{v}^0$ meets the problem $\mathrm{P_v}$, the vanishing of $\mathbf{v}^0(0) - \mathbf{v}_0$ and account of the boundary conditions (5.10.3) yield

$$\Theta_y(\mathbf{v}^0+\mathbf{h}) - \Theta(\mathbf{v}^0) = \tfrac{1}{2}\left[\rho(\dot{\mathbf{h}}, \mathbf{h})_y + (\mathbf{N}*\nabla\mathbf{h}, \nabla\mathbf{h})_y + \rho_0 \int_0^\infty y(t) \int_\Omega \mathbf{h}(0)\cdot\mathbf{h}(t)\, dx\, dt\right].$$

Hence, by the choice of the weight function y and the definition of $\mathbf{N}$, we obtain

$$\begin{aligned}&\Theta_y(\mathbf{v}^0 + \mathbf{h}) - \Theta_y(\mathbf{v}^0) = \\ &\tfrac{1}{2} \int_0^\infty Y(\alpha) \int_\Omega \Big[\rho_0 \alpha |\mathbf{h}_L(\alpha)|^2 + 2\,\mu_L(\alpha)|\mathrm{sym}\, \nabla \overset{\circ}{\mathbf{h}}_L(\alpha)|^2 \\ &\qquad + \Big(\kappa_L(\alpha) + \frac{\Pi_\rho \rho_0}{\alpha}\Big)|\nabla \cdot \mathbf{h}_L(\alpha)|^2\Big]\, dx\, d\alpha.\end{aligned}$$

Since $\Pi_\rho > 0$ and $\mu_L, \kappa_L > 0$, we conclude that $\Theta_y(\mathbf{v}^0 + \mathbf{h}) - \Theta_y(\mathbf{v}^0) \geq 0$ and equality holds only at $\mathbf{h} = 0$, that is, at $\mathbf{v} = \mathbf{v}^0$. □

Now that a minimum property is established for the velocity field, we may ask about the correspondent behaviour of φ. It turns out that if $\mathbf{v}^0 \in \bar{\mathcal{Q}}_v$ meets the problem $\mathrm{P_v}$, then the pair $(\mathbf{v}^0, \varphi^0)$, where $\varphi^0(t) = \varphi_0 - \rho_0 \int_0^t (\nabla \cdot \mathbf{v}^0)(s)\, ds$, meets the problem $\mathrm{P_f}$. To prove the result we have only to show that φ^0 obeys the constraint $(5.10.12)_1$. By the definition of φ^0, integration over Ω and the divergence theorem yield

$$\int_\Omega \varphi(t)\, dx = \int_\Omega \varphi_0\, dx + \rho_0 \int_0^t \int_{\partial\Omega} \mathbf{v}^0(t) \cdot \mathbf{n}\, da\, dt.$$

Since $\mathbf{v}^0 \in \bar{\mathcal{Q}}_v$ the surface integral vanishes and then

$$\int_\Omega \varphi(t)\, dx = \int_\Omega \varphi_0\, dx.$$

Accordingly, if the initial datum φ_0 of $\mathrm{P_f}$ satisfies the constraint $(5.10.12)_1$, so does φ^0 for every $t \in \mathbb{R}^+$.

By paralleling Theorem 5.10.2 for incompressible fluids, we can show easily that the converse of Theorem 5.10.3 holds, too.

THEOREM 5.10.4. *If* $\mathbf{v}^0 \in \bar{\mathcal{Q}}_v$ *and*

$$\Theta_y(\mathbf{v}^0) \leq \Theta_y(\mathbf{v}), \qquad \mathbf{v} \in \bar{\mathcal{Q}}_v,$$

for every $Y \in C_0^\infty(\mathbb{R}^+, \mathbb{R}^+)$, *then* $\mathbf{v}^0$ *is the unique solution to the problem* $\mathrm{P_v}$.

Proof. Represent $\mathbf{v} \in \bar{\mathcal{Q}}_v$ in the form $\mathbf{v} = \mathbf{v}^0 + \eta\mathbf{h}$, where $\mathbf{h} \in \bar{\mathcal{Q}}_v$, $\eta \in \mathbb{R}$. After some rearrangements we arrive at

$$\begin{aligned}\Theta_y(\mathbf{v}^0 + \eta\mathbf{h}) - \Theta_y(\mathbf{v}^0) = \eta\Bigg\{ &(\rho_0 \dot{\mathbf{v}}^0 - \nabla \cdot (\mathbf{N} * \nabla\mathbf{v}^0) - \rho_0\mathbf{f}, \mathbf{h})_y \\ &+ \rho_0 \int_0^\infty y(t) \int_\Omega [\mathbf{v}^0(0) - \mathbf{v}_0] \cdot \mathbf{h}(t)\, dx\, dt \\ &+ \int_0^\infty \int_0^\infty y(t+\tau) \int_{\partial\Omega_T} [\mathbf{n}\,\mathbf{N} * \nabla\mathbf{v}^0 - \hat{\mathbf{t}}_0](t) \cdot \mathbf{h}(\tau)\, da\, d\tau\, dt \Bigg\} \\ &+ \tfrac{1}{2}\eta^2 \left\{ \rho_0(\dot{\mathbf{h}}, \mathbf{h})_y + (\mathbf{N} * \nabla\mathbf{h}, \mathbf{h})_y + \rho_0 \int_0^\infty y(t) \int_\Omega \mathbf{h}(0) \cdot \mathbf{h}(t)\, dx\, dt \right\}.\end{aligned}$$

As with Theorem 5.10.2, the coefficient of η^2 is strictly positive for every nonzero $\mathbf{h}$ and then the minimum of Θ_y on $\bar{\mathcal{Q}}_v$ is unique. The vanishing of the coefficient of η for every $\mathbf{h} \in \bar{\mathcal{Q}}_v$ and $Y \in C_0^\infty(\mathbb{R}^+, \mathbb{R}^+)$ yields

$$\rho_0\, \alpha\, \mathbf{v}^0{}_L = \nabla \cdot (\mathbf{N}_L \nabla\mathbf{v}^0{}_L) + \rho_0\mathbf{f}_L + \rho_0\mathbf{v}_0$$

on Ω and

$$(\hat{\mathbf{t}}_0)_L = (\mathbf{N}_L \nabla \mathbf{v}^0{}_L)\mathbf{n}$$

on $\partial\Omega_T$. The uniqueness of the Laplace transform implies the uniqueness of the solution to the problem $P_{\mathbf{v}}$.

Chapter 6

Wave Propagation

6.1. Discontinuity waves. In this section, we review general properties of discontinuity waves with a twofold purpose: first, to establish preliminary relations and properties for subsequent developments; second, to show how the relaxation function affects the wave behaviour. This is performed by disregarding boundary or initial value problems which, rather, will be the subject of the next chapter. As we shall see, once again the thermodynamic restrictions play a prominent role in the behaviour of the material.

A discontinuity wave is regarded as a singular surface.[34] Consider a time-dependent surface $\sigma(t)$ in the present placement $\mathcal{R}_t$, which is the common boundary of two regions $\mathcal{R}_t^+$ and $\mathcal{R}_t^-$, $\mathcal{R}_t^+ \cup \mathcal{R}_t^- = \mathcal{R}_t$. To fix ideas, we suppose that $\sigma(t)$ moves toward $\mathcal{R}_t^+$. Letting f be any function defined on $(\mathcal{R}_t \times \mathbb{R}) \setminus \sigma(t)$, denote by $f^+(\mathbf{x},t)$ and $f^-(\mathbf{x},t)$ the defined limit values of $f(\tilde{\mathbf{x}},t)$ as $\tilde{\mathbf{x}}$ approaches a point $\mathbf{x}$ on $\sigma(t)$ while remaining within $\mathcal{R}_t^+$ and $\mathcal{R}_t^-$, respectively. We denote by $\mathbf{n}$ the unit normal to σ, directed toward $\mathcal{R}_t^+$. The surface $\sigma(t)$ represents a discontinuity wave with respect to f if the jump $[f]$, defined as

$$[f] = f^- - f^+,$$

does not vanish on $\sigma(t)$. In this event, we say that f possesses a jump discontinuity.

The study of wave propagation rests upon some relations among discontinuities at σ. For the sake of generality, we allow σ to be a curved surface. Represent σ as

$$\mathbf{x} = \mathbf{p}(V^\Gamma, t)$$

where $V^\Gamma, \Gamma = 1, 2$, is a pair of surface parameters. The vectors

$$\mathbf{p}_\Gamma = \frac{\partial \mathbf{p}}{\partial V^\Gamma}$$

[34]The reader who has no familiarity with the subject may consult [122], §§175, 176.

are the covariant base vectors on σ along the line V^Γ. The components $a_{\Gamma\Delta}$ of the surface metric and the components $b_{\Gamma\Delta}$ of the second fundamental form of the surface are defined by

$$a_{\Gamma\Delta} = \mathbf{p}_\Gamma \cdot \mathbf{p}_\Delta, \qquad b_{\Gamma\Delta} = \mathbf{n} \cdot \mathbf{p}_{\Gamma\Delta}$$

where $\mathbf{p}_{\Gamma\Delta} = \partial^2 \mathbf{p}/\partial V^\Gamma \partial V^\Delta$. For any function f on $\mathcal{R} \times \mathbb{R}$ the rate of change of the jump $[f]$ obeys the kinematic condition of compatibility

$$\frac{d}{dt}[f] = [\dot{f}] + U\,\mathbf{n} \cdot [\nabla f]. \tag{6.1.1}$$

The derivative d/dt is called the displacement (Thomas) derivative and provides the time rate of change for an observer at rest with σ. The scalar U denotes the local speed of propagation, namely the normal speed of the surface σ relative to the material particles of the body instantaneously situated on σ. As a consequence, if $[f] = 0$ then

$$[\dot{f}] + U\,\mathbf{n} \cdot [\nabla f] = 0. \tag{6.1.1$'$}$$

On σ, $[f]$ and $[\nabla f]$ are related by the geometric condition of compatibility

$$[\nabla f] = B\,\mathbf{n} + a^{\Delta\Gamma}\mathbf{p}_\Delta [f]_{;\Gamma} \tag{6.1.2}$$

where $B = [df/dn]$ is the jump of the normal derivative and the symbol $_{;\Gamma}$ denotes the covariant derivative with respect to V^Γ. Of course, if $[f]$ is constant on σ then

$$[\nabla f] = B\,\mathbf{n}. \tag{6.1.2$'$}$$

Moreover, $[\nabla\nabla f]$ is expressed by the iterated geometric condition of compatibility

(6.1.3)
$$[\nabla\nabla f] = C\,\mathbf{n} \otimes \mathbf{n} + 2\,\mathrm{sym}(\mathbf{n} \otimes \mathbf{p}^\Gamma)(B_{;\Gamma} + b^\Lambda_\Gamma [f]_{;\Lambda}) + \mathbf{p}^\Gamma \otimes \mathbf{p}^\Lambda([f]_{;\Gamma\Lambda} - b_{\Gamma\Lambda} B),$$

where $C = [d^2 f/dn^2]$. By (6.1.2) and (6.1.1), upon replacing f with $\dot{f}$, we have the iterated kinematic condition of compatibility

$$[\nabla \dot{f}] = B'\mathbf{n} + a^{\Delta\Gamma}\mathbf{p}_\Delta [\dot{f}]_{;\Gamma}, \tag{6.1.4}$$

$$[\ddot{f}] = -U\,B' + \frac{d[\dot{f}]}{dt}, \tag{6.1.5}$$

where $B' = [\![d\dot{f}/dn]\!] = -U\,C + dB/dt + a^{\Delta\Gamma}\,U_{;\Delta}[\![f]\!]_{;\Gamma}$.

When we describe the discontinuity wave in the reference placement $\mathcal{R}$, we need the duals of the previous conditions of compatibility. The correspondence is established by letting

$$\nabla \to \frac{\partial}{\partial \mathbf{X}}, \qquad \mathbf{n} \to \mathbf{N}, \qquad U \to U_N,$$

$\mathbf{N}$ being the unit normal to the inverse image of σ and U_N the speed of propagation (in $\mathcal{R}$); it is

$$\mathbf{n} = \frac{\mathbf{F}^T\mathbf{N}}{|\mathbf{F}^T\mathbf{N}|}, \qquad U = |\mathbf{F}\mathbf{n}|U_N.$$

We are now in a position to examine the main properties of discontinuity waves. The structure of the theory of linear viscoelasticity suggests that we characterize them through the spatial representation σ of the singular surface. Then we consider waves of order $m \geq 2$ in the form of singular surfaces as follows.

(i) The displacement $\mathbf{u}$ on $\mathcal{R}_t \times \mathbb{R}$ is $m-1$ times continuously differentiable while the mth derivatives may possess jump discontinuities across σ.

(ii) The body force $\mathbf{b}$ and the density ρ are continuous and, if $m \geq 3$, $m-2$ times continuously differentiable. The body is supposed to be homogeneous, namely the relaxation function $\mathbf{G}$ is taken to be independent of $\mathbf{x} \in \mathcal{R}_t$.

Usually the waves of order $m = 2$ are called acceleration waves.

As to the (local) propagation speed we have the following result.[35]

THEOREM 6.1.1. *The speed of propagation U of a wave is independent of the order m of the wave and satisfies the eigenvalue problem*

$$(\mathbf{Q}(\mathbf{n}) - \rho U^2\mathbf{1})\mathbf{q} = 0, \qquad \mathbf{q} \in V, \tag{6.1.6}$$

where $\mathbf{Q}(\mathbf{n})$ is the acoustic tensor defined as

$$\mathbf{Q}(\mathbf{n})\mathbf{w} = [\mathbf{G}_0(\mathbf{w} \otimes \mathbf{n})]\mathbf{n}, \qquad \mathbf{w} \in V. \tag{6.1.7}$$

Proof. Let $m = 2$. By the equation of motion (1.4.6) we have

$$\rho[\![\ddot{\mathbf{u}}]\!] = [\![\nabla \cdot \mathbf{T}]\!].$$

Then, since $\mathbf{G}$ is independent of $\mathbf{x}$ and $[\![\dot{\mathbf{u}}]\!] = 0$, by (3.1.2) and (6.1.1′) we can write

$$\mathbf{G}_0[\![\dot{\mathbf{E}}]\!] + \left[\!\!\left[\int_0^\infty \mathbf{G}'(s)\,\dot{\mathbf{E}}(t-s)\,ds\right]\!\!\right] + \rho U[\![\ddot{\mathbf{u}}]\!] = 0.$$

[35] It was proved by Herrera and Gurtin [81] for $m = 2$ and by Fisher and Gurtin [62] for any order m.

By (6.1.1′) and (6.1.2′), with $f = \dot{\mathbf{u}}$, we find that

$$[\dot{\mathbf{E}}] = -\frac{1}{U}\mathrm{sym}(\mathbf{n}\otimes\mathbf{q})$$

where $\mathbf{q} = [\ddot{\mathbf{u}}]$ is the polarization of the wave. Substitution yields the Fresnel-Hadamard propagation condition (6.1.6).

Now let $m > 2$. Consider the $(m-2)$th time derivative of (1.4.6). Observe that $\mathbf{u}^{(m-1)} := d^{m-1}\mathbf{u}/dt^{m-1}$ is continuous across σ whence, for example,

$$[\mathbf{E}^{(m-1)}] = -\frac{1}{U}\mathrm{sym}(\mathbf{n}\otimes\mathbf{q})$$

where $\mathbf{q} = [\mathbf{u}^{(m)}]$. Substitution completes the proof. □

Observe that, by thermodynamics, $\mathbf{G}_0$ is symmetric (cf. (3.2.8)). Then $\mathbf{Q}$ too is symmetric and there exist three, not necessarily distinct, propagation speeds. Moreover, by (3.1.1) and (3.2.14) $\mathbf{G}_0$ is also strongly elliptic, namely

$$\mathbf{v}\cdot[\mathbf{G}_0(\mathbf{v}\otimes\mathbf{w})]\mathbf{w} > 0$$

for any nonzero vectors $\mathbf{v}, \mathbf{w}$. Then the propagation speeds are real and nonzero.

Obviously, the eigenvalues of $\mathbf{Q}(\mathbf{n})$ depend on $\mathbf{n}$ only. Then the propagation speed U is constant if $\mathbf{n}$ is constant. Consider plane waves of the form

$$\mathbf{u}(\mathbf{x}, t) = \bar{\mathbf{u}}(x, t)$$

where $x = \mathbf{x}\cdot\mathbf{n}$. In connection with the wave of order m, we define the induced jump γ as

$$\gamma = -\left[\frac{\partial^{m+1}\bar{\mathbf{u}}}{\partial x^{m+1}}\right](-U)^{m+1}$$

and let

$$\mathbf{q} = [\mathbf{u}^{(m)}] = \left[\frac{\partial^{m}\bar{\mathbf{u}}}{\partial x^{m}}\right](-U)^{m}$$

be the polarization. Let $\mathbf{G}_0'$ be symmetric and define the *attenuation tensor* $\mathbf{Q}'(\mathbf{n})$ by

(6.1.8) $$\mathbf{Q}'(\mathbf{n})\mathbf{w} = [\mathbf{G}_0'(\mathbf{w}\otimes\mathbf{n})]\mathbf{n}, \qquad \mathbf{w}\in V.$$

The symmetry of $\mathbf{G}_0'$ in Sym implies that of $\mathbf{Q}'$ in V. The following result is essential to the investigation of the attenuation of the polarization in time.

THEOREM 6.1.2. *The polarization* $\mathbf{q}$ *of a plane wave of order* m *which propagates in a homogeneous medium with speed* U *is governed by the differential equation*

(6.1.9) $$\mathbf{Q}'(\mathbf{n})\mathbf{q} - 2\rho U^2\frac{d\mathbf{q}}{dt} = \mathbf{Q}(\mathbf{n})\gamma - \rho U^2\gamma.$$

Proof. Let $m = 2$. Differentiate the equation of motion with respect to x. By (6.1.7) we can write

$$\rho\left[\frac{\partial \ddot{\mathbf{u}}}{\partial x}\right] = \mathbf{Q}(\mathbf{n})\left[\frac{\partial^3 \bar{\mathbf{u}}}{\partial x^3}\right] + \left[\frac{\partial}{\partial x}\int_0^\infty \nabla \cdot \mathbf{G}'(s)\frac{\partial \mathbf{u}}{\partial \mathbf{x}}(t-s)\,ds\right]. \tag{6.1.10}$$

Examine the last term. Because $\mathbf{G}'$ is independent of $\mathbf{x}$ and the integral is continuous across the wave, we have

$$\left[\frac{\partial}{\partial x}\int_0^\infty \nabla \cdot \mathbf{G}'(s)\frac{\partial \mathbf{u}}{\partial \mathbf{x}}(t-s)\,ds\right] = -\frac{1}{U}\left[\frac{\partial}{\partial t}\int_0^\infty \mathbf{G}'(s)\frac{\partial^2 \mathbf{u}}{\partial \mathbf{x}\partial \mathbf{x}}(t-s)\,ds\right].$$

Since $\partial^2\mathbf{u}/\partial\mathbf{x}\partial\mathbf{x}$ suffers a jump discontinuity at the wave, to evaluate the jump we apply a procedure elaborated by Morro [98 (see also [100], Chap. 13). We get

$$\begin{aligned}\left[\frac{\partial}{\partial t}\int_0^\infty \mathbf{G}'(s)\frac{\partial^2 \mathbf{u}}{\partial \mathbf{x}\partial \mathbf{x}}(t-s)\,ds\right] &= \lim_{\lambda\to 0_+}\frac{1}{\lambda}\int_0^\infty \mathbf{G}'(s)\left[\frac{\partial^2 \mathbf{u}}{\partial \mathbf{x}\partial \mathbf{x}}\right]\mathcal{U}(\lambda - s)\,ds \\ &= \mathbf{G}'_0\left[\frac{\partial^2 \mathbf{u}}{\partial \mathbf{x}\partial \mathbf{x}}\right]\end{aligned}$$

where $\mathcal{U}$ is Heaviside step function. On applying twice (6.1.1) and (6.1.2′) we arrive at

$$\left[\frac{\partial \ddot{\mathbf{u}}}{\partial x}\right] = -\frac{2}{U}\frac{d\mathbf{q}}{dt} - \frac{1}{U}\boldsymbol{\gamma}.$$

Substitution into (6.1.10) and use of (6.1.8) provide the desired result (6.1.9).

If $m > 2$ then we differentiate the equation of motion $m - 1$ times with respect to x. Then we can parallel step by step the procedure for $m = 2$ to get the result (6.1.9). □

Concerning (6.1.9) it is apparent that things are very simple when $\boldsymbol{\gamma}$ is such that

$$\mathbf{Q}(\mathbf{n})\boldsymbol{\gamma} = \rho U^2\boldsymbol{\gamma}.$$

When this is the case we say that the wave is *axially similar*. For example, in one-dimensional bodies $\mathbf{Q}$, $\boldsymbol{\gamma}$, and $\mathbf{q}$ reduce to real numbers and then all waves are axially similar. Really, such waves are not confined to the one-dimensional case. As we show in a moment, waves in isotropic solids are axially similar.

The evolution of axially similar waves is governed by the following theorem.

THEOREM 6.1.3. *At a plane, axially similar wave of order m, propagating in the direction $\mathbf{n}$ with speed U, the polarization $\mathbf{q}$ is given by*

$$\mathbf{q}(t) = \exp\left(\frac{t}{2\,\rho U^2}\mathbf{Q}'\right)\mathbf{q}(0). \tag{6.1.11}$$

Proof. The result (6.1.11) follows at once from (6.1.9) and the observation that the right-hand side vanishes. Incidentally, the symmetry of $\mathbf{Q}'$ allows $\mathbf{q}$ to be written as

$$\mathbf{q}(t) = \sum_{i=1}^{3} \alpha_i \mathbf{a}_i \exp(\mu_i t)$$

where $2\mu_i \rho U^2$ and $\mathbf{a}_i$ are the ith eigenvalue and eigenvector of $\mathbf{Q}'$ while the coefficients α_i are determined by the initial conditions. □

The result (6.1.11) emphasizes the central role of $\mathbf{Q}'$ in the evolution of $\mathbf{q}$. By the thermodynamic inequality (3.2.13), $\mathbf{G}_0'$ is negative semidefinite in Sym. Then it follows from (6.1.7) that $\mathbf{Q}'$ is negative semidefinite in V, i.e.,

$$\mathbf{w} \cdot \mathbf{Q}' \mathbf{w} \leq 0$$

for any nonzero $\mathbf{w} \in V$. Hence it follows that the eigenvalues μ_i are negative and $\mathbf{q}$ decays in time.

To complete the scheme, consider the case $\mathbf{Q}(\mathbf{n})\gamma \neq \rho U^2 \gamma$; the corresponding waves are called *axially dissimilar.* For such waves the time evolution of $\mathbf{q}$ is given by a theorem which is due to Fisher [61].

THEOREM 6.1.4. *At a plane, axially dissimilar wave of order m, propagating in the direction* $\mathbf{n}$ *with speed* U*, the time dependence of the polarization* $\mathbf{q}$ *is given as follows.*

(i) *If* ρU^2 *is a distinct eigenvalue of* $\mathbf{Q}$*, with eigenvector* $\mathbf{e}$*, then*

$$\mathbf{q}(t) = \alpha\, \mathbf{e} \exp(\zeta t), \qquad \zeta = \frac{\mathbf{e} \cdot \mathbf{Q}' \mathbf{e}}{2\rho U^2}. \tag{6.1.12}$$

(ii) *If* ρU^2 *is a double eigenvalue of* $\mathbf{Q}$ *then*

$$\mathbf{q}(t) = \sum_{\nu=1}^{2} \alpha_\nu \mathbf{a}_\nu \exp(\zeta_\nu t) \tag{6.1.13}$$

where $2\zeta_\nu \rho U^2$ *is the eigenvalue corresponding to eigenvector* $\mathbf{a}_\nu$ *of the reduced attenuation matrix* $Q_{\alpha\beta} = \mathbf{e}_\alpha \cdot \mathbf{Q}' \mathbf{e}_\beta$, $\alpha, \beta = 1, 2$, *with* $\mathbf{e}_1, \mathbf{e}_2$ *two orthonormal basis vectors in the principal plane of* $\mathbf{Q}$.

Proof. If ρU^2 is a distinct eigenvalue of $\mathbf{Q}$, then there exists only one unit vector $\mathbf{e}$ such that $\mathbf{Q}\mathbf{e} = \rho U^2 \mathbf{e}$. Then

$$\mathbf{e} \cdot (\mathbf{Q}\gamma - \rho U^2 \gamma) = 0$$

and, by (6.1.9),

$$\left(\mathbf{Q}' \mathbf{q} - 2\rho U^2 \frac{d\mathbf{q}}{dt} \right) \cdot \mathbf{e} = 0.$$

If the polarization $\mathbf{q}$ is parallel to $\mathbf{e}$, $\mathbf{q}(t) = \beta(t)\mathbf{e}$, then

$$(\mathbf{e} \cdot \mathbf{Q}'\mathbf{e})\beta - 2\,\rho\,U^2 \frac{d\beta}{dt} = 0$$

and (6.1.12) follows.

If ρU^2 is a double eigenvalue of $\mathbf{Q}$, then there are two orthogonal unit vectors $\mathbf{e}_1, \mathbf{e}_2$ such that each eigenvector $\mathbf{b}$ of $\mathbf{Q}$, with eigenvalue ρU^2, can be written as $\mathbf{b} = b_\gamma \mathbf{e}_\gamma$. Again by (6.1.9) we have

$$\left(\mathbf{Q}'\mathbf{e}_\alpha b_\alpha - 2\,\rho\,U^2 \mathbf{e}_\alpha \frac{db_\alpha}{dt}\right) \cdot \mathbf{e}_\gamma = 0.$$

Therefore, letting $Q'_{\alpha\beta} = \mathbf{e}_\beta \cdot \mathbf{Q}'\mathbf{e}_\alpha$, we obtain

$$Q'_{\alpha\beta} b_\alpha - 2\,\rho\,U^2 \frac{db_\beta}{dt} = 0$$

whence (6.1.13).

Of course, if ρU^2 is a triple eigenvalue of $\mathbf{Q}$, then any vector $\boldsymbol{\gamma}$ is an eigenvector of $\mathbf{Q}$ and the corresponding waves are axially similar. □

Also for later reference, it is of interest to consider the particular case of isotropic solids for which (3.1.2) can be written as

(6.1.14)

$$\mathbf{T}(t) = 2\mu_0 \mathbf{E}(t) + \lambda_0 (\mathrm{tr}\,\mathbf{E})(t)\,\mathbf{1} + \int_0^\infty [2\mu'(s)\,\mathbf{E}(t-s) + \lambda'(s)(\mathrm{tr}\,\mathbf{E})(t-s)\,\mathbf{1}]ds$$

where $\mu(s)$ is the relaxation modulus in shear and $\lambda(s)$ is related to the relaxation modulus in dilation, $\kappa(s)$, by

$$\lambda = \kappa - \tfrac{2}{3}\mu.$$

By (6.1.7) we have

$$\mathbf{Q}(\mathbf{n})\,\mathbf{w} = \mu_0 \mathbf{w} + (\mu_0 + \lambda_0)(\mathbf{w} \cdot \mathbf{n})\,\mathbf{n}$$

and then $\mathbf{n}$ is an eigenvector of $\mathbf{Q}$, with eigenvalue $2\mu_0 + \lambda_0$, while any vector orthogonal to $\mathbf{n}$ is an eigenvector of $\mathbf{Q}$ with (double) eigenvalue μ_0. Accordingly, it follows from (6.1.4) that *longitudinal waves,* for which

$$\mathbf{n} \times \mathbf{q} = 0,$$

can exist with propagation speed

$$U_L = \sqrt{\frac{2\mu_0 + \lambda_0}{\rho}} \tag{6.1.15}$$

and *transverse waves,* for which

$$\mathbf{n} \cdot \mathbf{q} = 0,$$

can exist with propagation speed

$$U_T = \sqrt{\frac{\mu_0}{\rho}}. \tag{6.1.16}$$

Incidentally, both $\mu_0 > 0$ and $2\mu_0 + \lambda_0 > 0$ are guaranteed by the positive definiteness of $\mathbf{G}_0$. As for $\mathbf{Q}'$, we have

$$\mathbf{Q}'(\mathbf{n})\mathbf{w} = \mu_0' \, \mathbf{w} + (\mu_0' + \lambda_0')(\mathbf{w} \cdot \mathbf{n})\mathbf{n}$$

where $\mu_0' = \mu'(0)$, $\lambda_0' = \lambda'(0)$. It is evident that $\mathbf{Q}$ and $\mathbf{Q}'$ commute and have the same characteristic manifolds; the eigenvectors are the normal $\mathbf{n}$ (with distinct eigenvalue) and any pair of vectors orthogonal to $\mathbf{n}$ (and each other). This implies that, for isotropic bodies, all waves are axially similar. Then Theorem 6.1.3 applies to give the following result.

COROLLARY 6.1.1. *Plane waves of order m propagating in an isotropic viscoelastic body are either longitudinal or transverse.*

(i) *At longitudinal waves* $\boldsymbol{\gamma} \times \mathbf{q} = 0$ *and*

$$\mathbf{q}(t) = q_0 \mathbf{n} \exp\left(\frac{\lambda_0' + 2\mu_0'}{2(\lambda_0 + 2\mu_0)} t\right). \tag{6.1.17}$$

(ii) *At transverse waves* $\mathbf{q}$ *and* $\boldsymbol{\gamma}$ *are both orthogonal to* $\mathbf{n}$ *and*

$$\mathbf{q}(t) = q_0 \boldsymbol{\tau} \exp\left(\frac{\mu_0'}{2\mu_0} t\right), \qquad \boldsymbol{\tau} \cdot \mathbf{n} = 0. \tag{6.1.18}$$

The negative definiteness of $\mathbf{G}_0'$, and then of $\mathbf{Q}'$, implies that both μ_0' and $\lambda_0' + 2\mu_0'$ are negative. Then the thermodynamic requirements on the viscoelastic model amount to the decay of the polarization $\mathbf{q}$ in time.

6.2. Curved shock waves and induced discontinuities. The previous, preliminary analysis of waves in viscoelastic solids is now generalized so as to account for two features which permit a fruitful connection between wave behaviour and constitutive properties of the material. Here we consider shock waves in isotropic solids. The generalization consists in letting the discontinuity surface be curved and allowing also for the waves of higher order arising in the region perturbed by the shock wave (induced discontinuities). This generalization has been performed by Chen [14] and involves quite cumbersome calculations. Here we summarize the main steps of the procedure and the pertinent results about the evolution of the shock and the induced discontinuities.

For technical reasons, it is convenient to adopt the material representation. We go back to $\mathbf{E} = \operatorname{sym}(\nabla_{\mathbf{X}}\mathbf{u})$ where $\nabla_{\mathbf{X}} = \partial/\partial\mathbf{X}$ and let (6.1.14) still hold. Owing to the linearity of the theory, the Piola-Kirchhoff stress tensor $\mathbf{S} = J\mathbf{T}(\mathbf{F}^{-1})^T$ is identified with $\mathbf{T}$. Even though the same symbols as before are used, the pertinent expressions in terms of $\mathbf{X}$ and t are understood.

Denote by Σ the discontinuity surface in $\mathcal{R}$. A shock wave is a discontinuity wave of order $m = 1$. Then $\dot{\mathbf{u}}$ is discontinuous at Σ and we define the polarization $\mathbf{q}$ as

$$\mathbf{q} = [\dot{\mathbf{u}}]. \tag{6.2.1}$$

By the dual of (6.1.1′), with $f = \mathbf{u}$, we have

$$[\nabla_{\mathbf{X}}\mathbf{u}] = -\frac{1}{U_N}\mathbf{N}\otimes\mathbf{q}. \tag{6.2.2}$$

Consider the balance of linear momentum, in integral form, namely

$$\frac{d}{dt}\int_{\mathcal{P}} \rho_0 \mathbf{v}\, dX = \int_{\partial\mathcal{P}} J\,\mathbf{T}(\mathbf{F}^{-1})^T\,\mathbf{N}\, dA$$

for any regular region $\mathcal{P} \in \mathcal{R}$. Then Kotchine's theorem yields

$$[\mathbf{T}]\mathbf{N} = -\rho_0\, U_N\, \mathbf{q}. \tag{6.2.3}$$

By (6.1.14) we have $[\mathbf{T}] = 2\,\mu_0[\mathbf{E}] + \lambda_0[\operatorname{tr}\mathbf{E}]\,\mathbf{1}$. Then by (6.2.1)–(6.2.3) we conclude that, for every direction of propagation $\mathbf{N}$, every shock wave is longitudinal or transverse and the corresponding wave speeds U_{NL}, U_{NT} take the values

$$U_{NL} = \sqrt{\frac{2\mu_0 + \lambda_0}{\rho_0}}, \qquad U_{NT} = \sqrt{\frac{\mu_0}{\rho_0}}$$

which, within the approximation adopted, coincide with the speeds for acceleration waves (6.1.15), (6.1.16).

To derive the differential equation for the evolution of the shock, we start from the jump relation induced by the equation of motion, i.e., $[\nabla\cdot\mathbf{T}] = \rho_0[\ddot{\mathbf{u}}]$. By the duals of (6.1.2) and (6.1.1), with f replaced by $\mathbf{T}$, we have

$$[\nabla_{\mathbf{X}}\cdot\mathbf{T}] = \left[\frac{d}{dN}\mathbf{T}\right]\mathbf{N} + a^{\Delta\Gamma}[\mathbf{T}]_{;\Gamma}\mathbf{p}_{\Delta},$$

$$\frac{d}{dt}[\mathbf{T}] = [\dot{\mathbf{T}}] + U_N\Big[\frac{d}{dN}\mathbf{T}\Big],$$

whence

$$\frac{d}{dt}([\mathbf{T}]\mathbf{N}) - [\dot{\mathbf{T}}]\mathbf{N} + U_N\, a^{\Delta\Gamma}[\mathbf{T}]_{;\Gamma}\mathbf{p}_\Delta \tag{6.2.4}$$

where use has been made of the fact that, by the constancy of U_N relative to V^Γ, $d\mathbf{N}/dt = 0$. Incidentally, since

$$\mathbf{p}_{\Gamma\Delta} = b_{\Gamma\Delta}\,\mathbf{N}, \qquad \mathbf{N}_{;\Gamma} = -b_\Gamma^\Delta\,\mathbf{p}_\Delta,$$

by the duals of (6.1.4) and (6.1.5) we obtain

$$[\nabla_{\mathbf{X}}\dot{\mathbf{u}}] = -U_N\,\mathbf{N}\otimes\mathbf{c} - \frac{1}{U_N}\mathbf{N}\otimes\frac{d\mathbf{q}}{dt} + a^{\Delta\Gamma}\mathbf{p}_\Delta\otimes\mathbf{q}_{;\Gamma}, \tag{6.2.5}$$

$$[\ddot{\mathbf{u}}] = U_N^2\mathbf{c} + 2\frac{d\mathbf{q}}{dt}, \tag{6.2.6}$$

where $\mathbf{c} = \big[d^2\mathbf{u}/dN^2\big]$. Moreover, by (6.2.3), (6.1.14), and (6.2.2), along with (6.2.5) and (6.2.6), we have

$$\frac{d}{dt}([\mathbf{T}]\mathbf{N}) = -\rho_0\,U_N\,\frac{d\mathbf{q}}{dt}, \tag{6.2.7}$$

$$\begin{aligned}
-[\dot{\mathbf{T}}]\mathbf{N} =& U_N(\lambda_0+\mu_0)(\mathbf{c}\cdot\mathbf{N})\mathbf{N} + U_N\mu_0\mathbf{c} + \frac{1}{U_N}(\lambda_0+\mu_0)\Big(\frac{d\mathbf{q}}{dt}\cdot\mathbf{N}\Big)\mathbf{N}\\
&+\frac{1}{U_N}\mu_0\frac{d\mathbf{q}}{dt} - \lambda_0 a^{\Delta\Gamma}(\mathbf{p}_\Delta\cdot\mathbf{q}_{;\Gamma})\mathbf{N} - \mu_0 a^{\Delta\Gamma}(\mathbf{q}_{;\Gamma}\cdot\mathbf{N})\mathbf{p}_\Delta\\
&+\frac{1}{U_N}(\lambda_0'+\mu_0')(\mathbf{q}\cdot\mathbf{N})\mathbf{N} + \frac{1}{U_N}\mu_0'\mathbf{q},
\end{aligned} \tag{6.2.8}$$

$$U_N[\mathbf{T}]_{;\Gamma}\,\mathbf{p}_\Delta = -\lambda_0 a^{\Delta\Gamma}\mathbf{p}_\Delta(\mathbf{q}\cdot\mathbf{N})_{;\Gamma} - 2\,\mu_0 a^{\Delta\Gamma}(\mathrm{sym}(\mathbf{q}\otimes\mathbf{N}))_{;\Gamma}\,\mathbf{p}_\Delta. \tag{6.2.9}$$

THEOREM 6.2.1. *The evolution of the polarization* $\mathbf{q}$ *of a longitudinal shock wave satisfies the differential equation*

$$2\,\rho_0\,U_{NL}\frac{dq}{dt} = (\lambda_0 + 2\,\mu_0)b_\Gamma^\Gamma q + \frac{\lambda_0' + 2\,\mu_0'}{U_{NL}}q. \tag{6.2.10}$$

Proof. For longitudinal shock waves $\mathbf{q} = q\,\mathbf{N}$ and then we have

$$\mathbf{p}_\Delta\cdot\mathbf{q} = 0,$$

whence

$$\mathbf{p}_\Delta \cdot \mathbf{q}_{;\Gamma} = -b_{\Delta\Gamma}\, q, \tag{6.2.11}$$

$$(\mathbf{q} \otimes \mathbf{N})_{;\Gamma}\mathbf{p}_\Delta = (\mathbf{N} \otimes \mathbf{q})_{;\Gamma}\mathbf{p}_\Delta = -b_{\Delta\Gamma}\mathbf{q}. \tag{6.2.12}$$

The inner product of (6.2.4) with $\mathbf{N}$ and use of (6.2.7)–(6.2.9) and (6.2.11), (6.2.12) yields the result (6.2.10) for the shock amplitude q. □

Equation (6.2.10) shows two contributions to the evolution of q. One is due to the form of the surface Σ which enters through the mean curvature b_Γ^Γ. If the perturbed region $\mathcal{R}^-$ is convex, then $b_\Gamma^\Gamma < 0$ and the amplitude q tends to decay. The opposite happens if $\mathcal{R}^+$ is convex. The other contribution is due to the memory of the material through the initial values λ_0' and μ_0' of the Boltzmann functions. Because $\lambda_0' + \mu_0' \leq 0$, it always makes the amplitude to decay as the shock propagates. For plane shocks, i.e., $b_\Gamma^\Gamma = 0$ at σ, (6.2.12) reduces to (6.1.17), which was to be expected by the linearity of the theory.

THEOREM 6.2.2. *The evolution of* $q^2 = \mathbf{q} \cdot \mathbf{q}$ *at a transverse shock is governed by the differential equation*

$$\rho_0\, U_{NT} \frac{dq^2}{dt} = \mu_0 b_\Gamma^\Gamma\, q^2 + \frac{\mu_0'}{U_{NT}} q^2. \tag{6.2.13}$$

Proof. For transverse shock waves, i.e., $\mathbf{q} \cdot \mathbf{N} = 0$, the polarization $\mathbf{q}$ is a surface vector with Cartesian and surface components q_i and q^Γ related by

$$q_i = p_{i;\Gamma} q^\Gamma, \qquad q^\Gamma = q_i p_{i;}{}^\Gamma.$$

By the constancy of U_N we have

$$\frac{d\mathbf{p}_\Gamma}{dt} = -U_N\, b_\Gamma^\Delta\, \mathbf{p}_\Delta, \qquad \frac{d\mathbf{p}^\Gamma}{dt} = U_N\, b^{\Gamma\Delta}\, \mathbf{p}_\Delta.$$

By the transversality of the shock it follows that

$$\mathbf{q}_{;\Gamma} \cdot \mathbf{N} = -\mathbf{q} \cdot \mathbf{N}_{;\Gamma} = b_{\Gamma\Delta} q^\Delta$$

and

$$(\mathbf{q} \otimes \mathbf{N})_{;\Gamma}\mathbf{p}_\Delta = \mathbf{q}(\mathbf{N}_{;\Gamma} \cdot \mathbf{p}_\Delta) = -b_{\Gamma\Delta}\mathbf{q},$$

$$(\mathbf{N} \otimes \mathbf{q})_{;\Gamma}\, \mathbf{p}_\Delta = (\mathbf{p}_\Delta \cdot \mathbf{q}_{;\Gamma})\mathbf{N} - b_\Gamma^\Lambda \mathbf{p}_\Lambda q_\Delta.$$

Then the inner product of (6.2.4) with $\mathbf{p}^\Lambda$ and multiplication by q_Λ yields (6.2.13). □

Similar results hold for the components q_i and q^Λ of $\mathbf{q}$.

The evolution equation (6.2.13) shows a behaviour strictly analogous to that of longitudinal shocks. In particular, the convexity of the perturbed region results in a decay of the shock. The negativity of μ_0', required by thermodynamics, makes the shock to decay.

The behaviour of induced discontinuities[36] in the region perturbed by the shock is determined by starting from the jump relation corresponding to the time derivative of the equation of motion (1.4.6), namely

$$[\nabla_{\mathbf{X}} \cdot \dot{\mathbf{T}}] = \rho_0[\dddot{\mathbf{u}}],$$

whence

$$\frac{d}{dt}([\dot{\mathbf{T}}]\mathbf{N}) - [\ddot{\mathbf{T}}]\mathbf{N} + U_N\, \mathbf{p}^\Gamma[\dot{\mathbf{T}}]_{;\Gamma} = \rho_0\, U_N[\dddot{\mathbf{u}}].$$

Then the procedure is based on proficient applications of the geometric and kinematic conditions of compatibility (see [14] for the detailed proof). Let $[d\dot{\mathbf{u}}/dN]$ be the polarization of the induced discontinuity. For a longitudinal shock wave, we let $q_{;\Gamma} = 0$ and then the pertinent discontinuity is the normal component $\mathbf{N} \cdot [d\dot{\mathbf{u}}/dN]$. It turns out that

$$\begin{aligned} 2\,\rho_0\, U_{NL}\, \frac{d}{dt}\Big[\mathbf{N} \cdot \frac{d\dot{\mathbf{u}}}{dN}\Big] =& (\lambda_0 + 2\,\mu_0) b_\Lambda^\Lambda \Big[\mathbf{N} \cdot \frac{d\dot{\mathbf{u}}}{dN}\Big] + \frac{\lambda_0' + 2\,\mu_0'}{U_{NL}} \Big[\mathbf{N} \cdot \frac{d\dot{\mathbf{u}}}{dN}\Big] \\ &+ \frac{3(\lambda_0 + 2\,\mu_0)}{2} b_\Delta^\Lambda b_\Lambda^\Delta q + \frac{\lambda_0 + 2\,\mu_0}{4} (b_\Lambda^\Lambda)^2 q \\ &+ \frac{3(\lambda_0' + 2\,\mu_0')^2}{4\,\rho\, U_{NL}^4} q - \frac{\lambda_0'' + 2\,\mu_0''}{U_{NL}^2} q. \end{aligned} \tag{6.2.14}$$

This result shows the effects of the curvature of the surface Σ and those of the memory of the material. In particular, the last term is a qualitatively new memory effect in that it involves the second-order derivatives λ_0'', μ_0'' of $\lambda(s)$, $\mu(s)$ at $s = 0$.

For a transverse shock wave the pertinent discontinuity is the tangential part of $[d\dot{\mathbf{u}}/dN]$, namely $\mathbf{p}^\Gamma \cdot [d\dot{\mathbf{u}}/dN]$. We let $q^\Gamma{}_{;\Delta} = 0$, which amounts to $\mathbf{N} \cdot [d\dot{\mathbf{u}}/dN] = 0$. The inner product of (6.2.4) with $\mathbf{p}^\Gamma$ and a careful application of the compatibility conditions lead to the evolution equation for the Γ-component of the induced discontinuity in the form

$$\begin{aligned} 2\,\rho_0\, U_{NT} \frac{d}{dt}\Big[\mathbf{p}^\Gamma \cdot \frac{d\dot{\mathbf{u}}}{dN}\Big] =& \mu_0 b_\Lambda^\Lambda \Big[\mathbf{p}^\Gamma \cdot \frac{d\dot{\mathbf{u}}}{dN}\Big] + 2\,\mu_0 b_\Delta^\Gamma \Big[\mathbf{p}^\Delta \cdot \frac{d\dot{\mathbf{u}}}{dN}\Big] + \frac{\mu_0'}{U_{NT}} \Big[\mathbf{p}^\Gamma \cdot \frac{d\dot{\mathbf{u}}}{dN}\Big] \\ &+ \mu_0 b_\Lambda^\Lambda b_\Delta^\Gamma q^\Delta + \frac{\mu_0}{2} b_\Delta^\Lambda b_\Lambda^\Delta q^\Gamma + \frac{\mu_0}{4} (b_\Lambda^\Lambda)^2 q^\Gamma + \frac{3(\mu_0')^2}{4\rho U_{NT}^4} q^\Gamma - \frac{\mu_0''}{U_{NT}^2} q^\Gamma. \end{aligned} \tag{6.2.15}$$

---

[36] Here they are waves of order 2.

A close connection between the behaviour of the induced discontinuities and the constitutive properties of the material is established in [15] by disregarding the geometric effects. Letting Σ be a plane surface, we can write (6.2.14) and (6.2.15) in the particular forms

(6.2.14′)
$$2\,\rho_0\,U_{NL}\,\frac{d}{dt}\Big[\mathbf{N}\cdot\frac{d\dot{\mathbf{u}}}{dN}\Big] = \frac{\lambda_0'+2\,\mu_0'}{U_{NL}}\Big[\mathbf{N}\cdot\frac{d\dot{\mathbf{u}}}{dN}\Big] + \frac{3(\lambda_0'+2\,\mu_0')^2}{4\,\rho_0\,U_{NL}^4}q - \frac{\lambda_0''+2\,\mu_0''}{U_{NL}^2}q,$$

(6.2.15′)
$$2\,\rho_0\,U_{NT}\frac{d}{dt}\Big[\mathbf{p}^\Gamma\cdot\frac{d\dot{\mathbf{u}}}{dN}\Big] = \frac{\mu_0'}{U_{NT}}\Big[\mathbf{p}^\Gamma\cdot\frac{d\dot{\mathbf{u}}}{dN}\Big] + \frac{3(\mu_0')^2}{4\rho_0 U_{NT}^4}q^\Gamma - \frac{\mu_0''}{U_{NT}^2}q^\Gamma.$$

Consider a plane acceleration wave preceded by plane, longitudinal shock wave. Denote by

$$w = \Big[\mathbf{N}\cdot\frac{d\dot{\mathbf{u}}}{dn}\Big]$$

the amplitude of the induced discontinuity and take the initial conditions for q and w as

(6.2.16)
$$q(0) = q_0, \qquad w(0) = w_0.$$

THEOREM 6.2.3. *The discontinuity induced by a longitudinal shock wave, with the initial conditions* (6.2.16), *is given by the function*

(6.2.17)
$$w(t) = w_0 \exp\left(\frac{\lambda_0'+2\,\mu_0'}{2\,\rho_0\,U_{NL}^2}t\right) + q_0\left[\frac{3(\lambda_0'+2\,\mu_0')^2}{8\,\rho_0^2\,U_{NL}^5} - \frac{\lambda_0''+2\,\mu_0''}{2\,\rho_0\,U_{NL}^3}\right] t\,\exp\left(\frac{\lambda_0'+2\,\mu_0'}{2\,\rho_0\,U_{NL}^2}t\right), \qquad t\in\mathbb{R}^+.$$

Proof. By the evolution equation for the induced discontinuity (6.2.14′), we obtain

(6.2.18)
$$w(t) = w_0 \exp\left(\frac{\lambda_0'+2\,\mu_0'}{2\,\rho_0\,U_{NL}^2}t\right) + \left[\frac{3(\lambda_0'+2\,\mu_0')^2}{8\,\rho_0^2\,U_{NL}^5} - \frac{\lambda_0''+2\,\mu_0''}{2\,\rho_0\,U_{NL}^3}\right]\int_0^t \exp\left[\frac{\lambda_0'+2\,\mu_0'}{2\,\rho_0\,U_{NL}^2}(t-\tau)\right] q(\tau)\,d\tau.$$

The function $q(\tau)$ is solution, in $\mathbb{R}^+$, of the evolution equation for the longitudinal shock (6.2.10), with $b_\Lambda^\Lambda = 0$, and then we have

$$q(\tau) = q_0 \exp\Big(\frac{\lambda_0'+2\,\mu_0'}{2\,\rho_0\,U_{NL}^2}\tau\Big).$$

Substitution in (6.2.18) provides the result (6.2.17). □

By the same token, we can prove the analogous result for transverse shocks. Let

$$w^\Gamma = \left[\mathbf{p}^\Gamma \cdot \frac{d\dot{\mathbf{u}}}{dN}\right]$$

and

$$q^\Gamma(0) = q_0^\Gamma, \qquad w^\Gamma(0) = w_0^\Gamma. \tag{6.2.19}$$

THEOREM 6.2.4. *The discontinuity induced by a longitudinal shock wave with the initial conditions* (6.2.19) *is given by the function*

$$\begin{aligned} w^\Gamma(t) = w_0^\Gamma \exp\Big(\frac{\mu_0'}{2\,\rho_0\,U_{NT}^2}t\Big) \\ + q_0^\Gamma\Big[\frac{3(\mu_0')^2}{8\rho_0^2 U_{NT}^5} - \frac{\mu_0''}{2\,\rho_0\,U_{NT}^3}\Big]\,t\,\exp\Big(\frac{\mu_0'}{2\,\rho_0\,U_{NT}^2}t\Big), \qquad t \in \mathbb{R}^+. \end{aligned} \tag{6.2.20}$$

Incidentally, the choice of $t = 0$ as the initial time is a matter of formal convenience which is allowed by the solutions being time invariant in form. For instance, for any time t_0 we can write (6.2.17) as

$$\begin{aligned} w(t) =& w(t_0)\exp\Big[\frac{\lambda_0' + 2\,\mu_0'}{2\,\rho_0\,U_{NL}^2}(t-t_0)\Big] \\ &+ q(t_0)\Big[\frac{3(\lambda_0' + 2\,\mu_0')^2}{8\,\rho_0^2\,U_{NL}^5} - \frac{\lambda_0'' + 2\,\mu_0''}{2\,\rho_0\,U_{NL}^3}\Big](t-t_0)\exp\Big[\frac{\lambda_0' + 2\,\mu_0'}{2\,\rho_0\,U_{NL}^2}(t-t_0)\Big], \end{aligned}$$

where $t \geq t_0$.

6.3. Wave propagation and thermodynamics with internal variables. It is of interest to examine the behaviour of the amplitudes of the induced discontinuities and to relate it with thermodynamic requirements. We discuss only longitudinal waves, described by (6.2.17); strictly analogous considerations hold for transverse waves, governed by (6.2.20).

The value $w(t)$ consists of two contributions. The first one is the effect of the initial value w_0 of the induced discontinuity while the second one is just the effect of the shock at all subsequent times. Since $\lambda_0' + 2\mu_0' \leq 0$, the first contribution monotonically decreases in time. The second one, however, initially increases and then monotonically decreases as $t > -2\,\rho_0\,U_{NL}^2/(\lambda_0' + 2\mu_0')$. As a whole, the discontinuity may have a more involved behaviour if the sign of w_0 differs from that of $q_0[3(\lambda_0'+2\,\mu_0')^2/8\,\rho_0^2\,U_{NL}^5 - (\lambda_0''+2\,\mu_0'')/2\,\rho_0\,U_{NL}^3]$. In this regard, observe that the sign of $\lambda_0'' + 2\,\mu_0''$ is fully unrestricted by thermodynamics though quite often the assumption $\lambda_0'' + 2\,\mu_0'' > 0$ is made.

If $\lambda'_0 + 2\,\mu'_0$ were positive, then the amplitude of the induced discontinuity would tend to infinity as $t \to \infty$. This shows that, as is the case for any wave evolution considered so far, only contradiction with thermodynamics would lead to instability in wave propagation, i.e., unbounded amplitudes. Of course, this is strictly true for linear viscoelasticity.

Yet there are responses of some energetic materials [73] which might suggest that, in a sense, we allow for the possibility $\lambda'(s)+2\,\mu'(s) > 0$ and/or $\mu'(s) > 0$ as s runs in a suitable interval. However, any operative approach to modelling material behaviour should involve a finite number of degrees of freedom. We know that a linear viscoelastic body may be viewed as a material with internal variables in which case the relaxation function is an exponential function. In this case, thermodynamics forces the model to satisfy $\mu'(s) < 0$, $\lambda'(s) + 2\,\mu'(s) < 0$, $s \in \mathbb{R}^+$. So any possibility of energetic behaviour seems to be confined to nonlinear models. Following [95], here we outline a procedure for setting up nonlinear models which are compatible with thermodynamics.

For formal simplicity, consider the one-dimensional case and let the state of the body be the pair (E, ξ), ξ being an internal variable. The response functions ψ, T, and $\dot{\xi}$, of E and ξ, are regarded as physically admissible if and only if they satisfy the Clausius-Duhem inequality

$$-\psi_E \dot{E} - \psi_\xi \dot{\xi} + T\,\dot{E} \geq 0 \tag{6.3.1}$$

for any process $P = \dot{E}$. Hence

$$T = \psi_E, \tag{6.3.2}$$

$$A\dot{\xi} \geq 0, \tag{6.3.3}$$

where $A = -\psi_\xi$. Any pair of functions ψ, $\dot{\xi}$ satisfying (6.3.3) determines a stress function T through (6.3.2), which obviously meets (6.3.1). The purpose here is to find one of such pairs. In this regard, assume that the change of variable $\xi \to A$ is allowed and that there exists a function $\mathcal{D}$ (often called a dissipation function) such that

$$\dot{\xi} = \mathcal{D}_A. \tag{6.3.4}$$

Then (6.3.3) becomes

$$A\,\mathcal{D}_A \geq 0. \tag{6.3.5}$$

Any variable A and function $\mathcal{D}$, such that (6.3.5) holds, determine a thermodynamically admissible, constitutive model.

By analogy with §3.1 let

$$T = \mu_0 E + F(E)\,\xi,$$

the function $F(E)$ being introduced to account for nonlinearities. By (6.3.2) it follows that

$$\psi = \tfrac{1}{2}\mu_0 E^2 + \mathcal{F}(E)\xi + \hat{\psi}(\xi)$$

where $\mathcal{F}$ is the integral of F. For simplicity, choose

$$\hat{\psi}(\xi) = \tfrac{1}{2}\upsilon\xi^2, \qquad \upsilon > 0.$$

Then

$$A = -(\mathcal{F}(E) + \upsilon\,\xi)$$

and the change of variable $\xi \to A$ is well defined. Finally, let

$$\mathcal{D} = \tfrac{1}{2}a\,A^2, \qquad a > 0,$$

which is the simplest way of satisfying (6.3.5). Of course, we may regard a as a function of E which plays the role of a parameter in the change $\sigma \to A$. By (6.3.4)

$$\dot{\xi} = -a(E)(\mathcal{F}(E) + \upsilon\xi)$$

whence

$$\xi(t) = -\int_{-\infty}^{t} a(E(\zeta))\,\mathcal{F}(E(\zeta))\,\exp\Big[-\upsilon\int_{\zeta}^{t} a(E(\nu))d\nu\Big]d\zeta.$$

Then we obtain the constitutive equation for T as

$$T(t) = \mu_0 E(t) - F(E(t))\int_{-\infty}^{t} a(E(\zeta))\mathcal{F}(E(\zeta))\,\exp\Big[-\upsilon\int_{\zeta}^{t} a(E(\nu))d\nu\Big]d\zeta$$

or, if a is constant,

$$T(t) = \mu_0 E(t) - a\,F(E(t))\int_0^{\infty} \exp(-a\,\upsilon\,s)\,\mathcal{F}(E(t-s))\,ds. \tag{6.3.6}$$

When F is also constant, (6.3.6) provides the usual relation of linear viscoelasticity with Boltzmann function

$$G'(s) = -a\,F^2\,\exp(-a\,\upsilon\,s).$$

Because $a > 0$, then, as it must be, $G' < 0$, namely the solid exhibits relaxation. If instead F depends on E, then, depending on the history E^t, we may have $F(E(t))\,\mathcal{F}(E(t-s)) < 0$ in a suitable interval for s.

More involved behaviours can be incorporated in the three-dimensional analogue. By following the same conceptual procedure, in [95] a stress-strain relation is derived in the form

$$
\begin{aligned}
\mathbf{T}(t) =& \mu_0 \overset{\circ}{\mathbf{E}} (t) + \beta_0 (\operatorname{tr} \mathbf{E})(t)\, \mathbf{1} \\
& - \left\{ F \overset{\circ}{\mathbf{E}} \cdot \int_{-\infty}^{t} \exp\left[- \int_{s}^{t} \upsilon\, a(\mathbf{E}(\zeta))\, d\zeta \right] (a\, \mathcal{F}\, \overset{\circ}{\mathbf{E}})(s)\, ds \right\} \mathbf{1} \\
& - \frac{H}{3} \int_{-\infty}^{t} \exp\left[- \int_{s}^{t} \chi\, a(\mathbf{E}(\zeta))\, d\zeta \right] (a\, \mathcal{H})(s)\, ds\, \mathbf{1} \\
& - \mathcal{F} \int_{-\infty}^{t} \exp\left[- \int_{s}^{t} \upsilon\, a(\mathbf{E}(\zeta))\, d\zeta \right] (a\, \mathcal{F}\, \overset{\circ}{\mathbf{E}})(s)\, ds
\end{aligned}
$$

where F, H are functions of tr$\mathbf{E}$ and $\mathcal{F}$, $\mathcal{H}$ are their integrals while υ, χ are positive constants.

This shows how the procedure may generate nonlinear stress-strain relations which are compatible with thermodynamics. Of course, once we can contrast these relations with precise experimental data, the procedure may lead to physically realistic models.

6.4. Time-harmonic waves. Generally wave propagation properties result in a remarkable information on the constitutive parameters of a model. This is especially true for time-harmonic wave propagation in that simple, analytical expressions can be derived which, further, are easily correlated with experiments.

In this section, we investigate the basic features of (single-frequency) time-harmonic waves. This is performed by following a recent research on the subject [12], [13] which seems to provide a more profitable way toward the understanding of prominent phenomena occurring at fluid-solid or solid-solid interfaces.

Back to the spatial description, assume that, for any $\mathbf{x} \in \Omega$, $\mathbf{u}(\mathbf{x}, \cdot) \in L^1(\mathbb{R})$. Then consider the Fourier transform

$$\mathbf{u}_F(\mathbf{x}, \omega) = \int_{-\infty}^{\infty} \exp(-i\omega\tau)\, \mathbf{u}(\mathbf{x}, \tau)\, d\tau$$

for any $\omega \in \mathbb{R}$. Since $\mathbf{u}$ is a real-valued vector, it follows that the complex conjugate $\mathbf{u}_F^*$ is related to $\mathbf{u}_F$ by

$$\mathbf{u}_F^*(\mathbf{x}, \omega) = \mathbf{u}_F(\mathbf{x}, -\omega). \tag{6.4.1}$$

The condition (6.4.1) in turn ensures that

$$\mathbf{u}(\mathbf{x}, t) = \frac{1}{2\pi} \int_{-\infty}^{\infty} \exp(i\omega t)\, \mathbf{u}_F(\mathbf{x}, \omega)\, d\omega \tag{6.4.2}$$

provides a real-valued field for any time $t \in \mathbb{R}$.

In fact, a physical wave is hardly a single-frequency oscillation. Yet, because of linearity we can evaluate the effects of single-frequency oscillations and then superposition in the form (6.4.2) yields the physical field. This allows us to consider the single-frequency displacement

$$\mathbf{u}(\mathbf{x}, t; \omega) = \mathbf{u}_F(\mathbf{x}, \omega) \exp(i\omega t)$$

whose values are complex vectors. Moreover we search for plane wave solutions. Within the large variety of notation, it seems more usual to write plane wave solutions as

$$\mathbf{u}(\mathbf{x}, t; \omega) = \mathbf{p} \exp[i(\omega t - \mathbf{k} \cdot \mathbf{x})] \tag{6.4.3}$$

where $\mathbf{p}$ is the polarization and $\mathbf{k}$ the wave number vector, both possibly dependent on the (angular) frequency ω. Whenever $\mathbf{p}$ and $\mathbf{k}$ are complex-valued, the wave (6.4.3) is said to be inhomogeneous. When needed we represent $\mathbf{p}$ and $\mathbf{k}$ as

$$\mathbf{p} = \mathbf{p}_1 + i\,\mathbf{p}_2, \qquad \mathbf{k} = \mathbf{k}_1 + i\,\mathbf{k}_2$$

where $\mathbf{p}_1, \mathbf{p}_2, \mathbf{k}_1$ and $\mathbf{k}_2$ are real-valued. Then the planes $\mathbf{k}_1 \cdot \mathbf{x} =$ constant are planes of constant phase and $\mathbf{k}_2 \cdot \mathbf{x} =$ constant are planes of constant amplitude. The quantity $c = |\omega|/|\mathbf{k}_1|$ is called phase speed.

Owing to (6.4.1), $\mathbf{p}$ and $\mathbf{k}$ are required to satisfy appropriate conditions under the frequency inversion $\omega \to -\omega$. By the arbitrariness of $\mathbf{x}$, the property $\mathbf{u}^*(\mathbf{x}, t; \omega) = \mathbf{u}(\mathbf{x}, t; -\omega)$ implies that

$$\mathbf{p}_1(\omega) = \mathbf{p}_1(-\omega), \qquad \mathbf{p}_2(\omega) = -\mathbf{p}_2(-\omega)$$

and

$$\mathbf{k}_1(\omega) = -\mathbf{k}_1(-\omega), \qquad \mathbf{k}_2(\omega) = \mathbf{k}_2(-\omega). \tag{6.4.4}$$

Possible inhomogeneous wave solutions are determined by requiring that the motion (6.4.3) satisfy the equation of motion (1.4.6). Really, we follow the customary, though often tacit, approximation that the effects of the body force are negligible. Accordingly, the equation of motion (1.4.6) and the constitutive equation (3.1.2) imply that the motion (6.4.3) is allowed if and only if the propagation condition

$$\rho\omega^2 \mathbf{p} = \hat{\mathbf{Q}}(\mathbf{k}, \omega)\mathbf{p} \tag{6.4.5}$$

holds with

$$\hat{\mathbf{Q}}(\mathbf{k}, \omega)\mathbf{p} := \mathbf{k}[(\mathbf{G}_0 + \mathbf{G}'_c(\omega) - i\mathbf{G}'_s(\omega))(\mathbf{k} \otimes \mathbf{p})];$$

$\hat{\mathbf{Q}}$ is the complex analogue of the acoustic tensor $\mathbf{Q}$ of §6.1 and the subscripts c and s denote, as usual, the half-range cosine and sine transform. So time-harmonic waves are allowed whenever the polarization $\mathbf{p}$ is an eigenvector of the complex-valued tensor $\hat{\mathbf{Q}}(\mathbf{k})$. In such a case, the propagation condition (6.4.5) provides the dispersion relation $\mathbf{k} = \mathbf{k}(\omega)$.

The following lemma gives a property of $\hat{\mathbf{Q}}$ induced by thermodynamics.

LEMMA 6.4.1. *The inequality* (3.2.11) *implies that*

$$\omega \mathbf{w} \cdot \Im \hat{\mathbf{Q}}(\mathbf{v}, \omega)\mathbf{w} > 0, \qquad \omega \in \mathbb{R} \setminus \{0\}, \tag{6.4.6}$$

holds for every pair of real vectors $\mathbf{v}$, $\mathbf{w}$.

Proof. If $\mathbf{v}$, $\mathbf{w}$ are real vectors, then

$$\mathbf{w} \cdot \Im \hat{\mathbf{Q}}(\mathbf{v}, \omega)\mathbf{w} = -[\mathbf{v} \otimes \mathbf{w}] \cdot \mathbf{G}'_s(\omega)[\mathbf{v} \otimes \mathbf{w}].$$

By (3.2.11) we obtain the result (6.4.6). □

Incidentally, in the one-dimensional case, (6.4.5) provides

$$\frac{\omega^2}{k^2} = \frac{G_0 + G'_c(\omega) - iG'_s(\omega)}{\rho}.$$

This corresponds to a wave speed $c = \omega / k_1$ given by

$$c^2 = \frac{2}{\rho} \frac{1}{\sqrt{(G_0 + G'_c)^2 + (G'_s)^2} + G_0 + G'_c}.$$

By (3.2.11), or (6.4.6), it follows that, for any nonzero frequency ω, the phase speed c is a nonzero, finite quantity. Really, by Lemma 4.3.1, even at $\omega = 0$ the right-hand side turns out to be a strictly positive, finite quantity; this simply means that k_1 vanishes as ω vanishes.

Detailed and suggestive results follow when the solid is isotropic; that is why this assumption is made for the rest of the chapter. On account of the constitutive equation (6.1.14), the propagation condition (6.4.5) simplifies to

$$\rho \omega^2 \mathbf{p} = (\hat{\lambda} + \hat{\mu})(\mathbf{k} \cdot \mathbf{p})\mathbf{k} + \hat{\mu}(\mathbf{k} \cdot \mathbf{k})\mathbf{p} \tag{6.4.7}$$

where

(6.4.8)

$$\hat{\lambda} = \lambda_0 + \int_0^\infty \lambda'(s) \exp(-i\omega s)\, ds, \qquad \hat{\mu} = \mu_0 + \int_0^\infty \mu'(s) \exp(-i\omega s)\, ds.$$

Incidentally, by (6.4.8) we have

$$\Im \hat{\mu} = -\mu'_s, \qquad \Im \hat{\lambda} = -\lambda'_s.$$

The next developments are based on two inequalities of thermodynamic character on $\hat{\mu}$ and $\hat{\lambda}$ which correspond to (6.4.6).

LEMMA 6.4.2. *The inequality* (3.2.11) *implies that*

$$\omega\,\Im\hat{\mu} > 0, \qquad \omega\,\Im(\hat{\lambda} + \tfrac{2}{3}\hat{\mu}) > 0, \qquad \omega \in \mathbb{R} \setminus \{0\}, \tag{6.4.9}$$

and then

$$\omega\Im\hat{\mu} > 0, \qquad \omega\Im(\hat{\mu}+\hat{\lambda}) > 0, \qquad \omega\Im(2\hat{\mu}+3\hat{\lambda}) > 0, \qquad \omega \in \mathbb{R}\setminus\{0\}. \tag{6.4.10}$$

Proof. Let $\mathbf{E} \in \mathrm{Sym}$ be real. For isotropic bodies,

$$\mathbf{E} \cdot \mathbf{G}'_s\,\mathbf{E} = 2\,\mu'_s\ \overset{\circ}{\mathbf{E}} \cdot \overset{\circ}{\mathbf{E}} + (\lambda'_s + \tfrac{2}{3}\mu'_s)(\operatorname{tr}\mathbf{E})^2.$$

Then (3.2.11) implies the inequalities (6.4.9). The trivial observation that $\hat{\lambda}+\hat{\mu} = (\hat{\lambda}+2\hat{\mu}/3)+\hat{\mu}/3$ and the analogous one for $\hat{\lambda}+2\hat{\mu}$ yield the inequalities (6.4.10). □

THEOREM 6.4.1. *Inhomogeneous wave solutions exist if and only if either*

$$\mathbf{k} \cdot \mathbf{p} = 0, \qquad \mathbf{k} \cdot \mathbf{k} = \frac{\rho\omega^2}{\hat{\mu}} \tag{6.4.11}$$

or

$$\mathbf{k} \cdot \mathbf{p} = 0, \qquad \mathbf{k} \cdot \mathbf{k} = \frac{\rho\omega^2}{2\hat{\mu} + \hat{\lambda}}. \tag{6.4.12}$$

In either case,

$$\omega\Im(\mathbf{k} \cdot \mathbf{k}) < 0, \qquad \omega \in \mathbb{R} \setminus \{0\}. \tag{6.4.13}$$

Proof. Though $\mathbf{p}$ and $\mathbf{k}$ are complex-valued, we can parallel the standard reasoning for the real case. Observe that, by (6.4.10), $\hat{\mu} \neq 0$ and $2\hat{\mu} + \hat{\lambda} \neq 0$. If $\mathbf{k} \cdot \mathbf{p} = 0$ then (6.4.7) holds with $\mathbf{k} \cdot \mathbf{k} = \rho\omega^2/\hat{\mu}$. If, instead, $\mathbf{k} \cdot \mathbf{p} \neq 0$ then $\mathbf{p}$ and $\mathbf{k}$ must be parallel and (6.4.7) holds with $\mathbf{k} \cdot \mathbf{k} = \rho\omega^2/(2\,\hat{\mu} + \hat{\lambda})$. Finally, by (6.4.9) we obtain (6.4.13). The "if" part of the proof is obvious. □

Quite naturally, the solutions (6.4.11) and (6.4.12) are called transverse and longitudinal waves. Here, however, the vectors $\mathbf{p}$, $\mathbf{k}$ are complex-valued and then orthogonality and parallelism have to be envisaged in that sense.

To find the explicit form of $\mathbf{k}$ for transverse and longitudinal waves, observe that, by (6.4.8),

$$\hat{\lambda}^*(\omega) = \hat{\lambda}(-\omega), \qquad \hat{\mu}^*(\omega) = \hat{\mu}(-\omega).$$

Then, letting A stand for $\hat{\mu}$ or $2\hat{\mu}+\hat{\lambda}$, the real part A_1 and the imaginary part A_2 satisfy

$$A_1(\omega)=A_1(-\omega), \qquad A_2(\omega)=-A_2(-\omega),$$

namely A_1 is even while A_2 is odd. Moreover, by the thermodynamic requirement (6.4.9) we have

$$\omega A_2(\omega)>0, \qquad \omega\in\mathbb{R}.$$

In both cases (6.4.11), (6.4.12), the moduli k_1 and k_2 of $\mathbf{k}_1$ and $\mathbf{k}_2$ satisfy the system of equations

(6.4.14) $$k_1^2-k_2^2=\frac{\rho\omega^2A_1}{A_1^2+A_2^2},$$

(6.4.15) $$2k_1k_2\cos\theta=-\frac{\rho\omega^2A_2}{A_1^2+A_2^2}$$

where θ is the angle between $\mathbf{k}_1$ and $\mathbf{k}_2$. The right-hand sides are given functions of the frequency ω. Then we can determine $k_1(\omega),k_2(\omega)$, the angle θ being regarded as a parameter. Indeed, θ itself depends on ω in that, by (6.4.15), it must switch from a strictly negative value when $\omega>0$ to a strictly positive value when $\omega<0$. Letting $\chi=1/\cos^2\theta$, we have

$$\cos\theta=-\frac{1}{\sqrt{\chi}}\operatorname{sgn}A_2.$$

A trivial calculation yields the following result.

THEOREM 6.4.2. *The solution to the system* (6.4.14)–(6.4.15) *is*

$$k_1=\sqrt{\frac{\rho\omega^2}{2}}\sqrt{\frac{A_1+\sqrt{A_1^2+\chi A_2^2}}{A_1^2+A_2^2}},$$

$$k_2=\sqrt{\frac{\rho\omega^2}{2}}\sqrt{\frac{-A_1+\sqrt{A_1^2+\chi A_2^2}}{A_1^2+A_2^2}}.$$

Remark 6.4.1. In accordance with (6.4.4), we may regard the unit vector of $\mathbf{k}_2$ as independent of ω. As the sign of ω changes, the unit vector of $\mathbf{k}_1$ reverses. The unit vector of $\mathbf{k}_1/\omega$ (slowness vector) is then the same for each nonzero frequency. This implies that $\mathbf{k}_2\cdot\mathbf{k}_1/\omega<0$, $\omega\neq0$, whereby any wave decays while propagating.

Remark 6.4.2. Young's modulus $\hat{E}$ for a viscoelastic (isotropic) solid is given by

$$\hat{E} = \frac{\hat{\mu}(3\hat{\lambda} + 2\hat{\mu})}{\hat{\lambda} + \hat{\mu}}.$$

The inequalities (6.4.10), which are placed by thermodynamics, guarantee that $\hat{E}$ never vanishes. Now, by repeating step by step Pochhammer's treatment for longitudinal waves in cylindrical bars, we might conclude that $\rho\omega^2/\hat{E}$ is the value of $\mathbf{k}\cdot\mathbf{k}$ at the lowest order of approximation in the radius. Then we can reasonably say that the inequalities (6.4.10) ensure also the existence of longitudinal waves in cylindrical viscoelastic bars.

It has been shown by Coleman and Gurtin [21] that, for one-dimensional waves, the limit values c_∞ and $k_{2\infty}$ as ω tends to infinity satisfy

$$c_\infty = \sqrt{\frac{G_0}{\rho}}, \qquad c_\infty k_{2\infty} = -\frac{G_0'}{2G_0};$$

the quantities c_∞ and $c_\infty k_{2\infty}$ are called *ultrasonic speed* and *ultrasonic attenuation*. In such a way they established an equivalence between ultrasonic speed and attenuation for time-harmonic waves and speed and attenuation for acceleration waves. It is of interest to evaluate the analogous limits in the three-dimensional case.

THEOREM 6.4.3. *If* $\mu', \lambda', \mu'', \lambda'' \in C(\mathbb{R}^+) \cap L^1(\mathbb{R}^+)$, *then the ultrasonic speed* c_∞ *and the ultrasonic attenuation* $c_\infty k_{2\infty}$ *are given by*

$$c_\infty = \sqrt{\frac{\mu_0}{\rho}}, \quad \sqrt{\frac{2\mu_0 + \lambda_0}{\rho}}, \tag{6.4.16}$$

$$c_\infty k_{2\infty} = \frac{\mu_0'\sqrt{\chi}}{2\mu_0}, \quad \frac{(2\mu_0' + \lambda_0')\sqrt{\chi}}{2(2\mu_0 + \lambda_0)}. \tag{6.4.17}$$

Proof. By definition

$$c = \sqrt{\frac{2(A_1^2 + A_2^2)}{\rho(A_1 + \sqrt{A_1^2 + A_2^2})}}.$$

Since $\mu', \lambda' \in C(\mathbb{R}^+) \cap L^1(\mathbb{R}^+)$ by Riemann-Lebesgue's lemma, we have, $A_2 \to 0$ and $A_1 \to \mu_0,\ 2\mu_0 + \lambda_0$ as $\omega \to \infty$. Hence we get the result (6.4.16).

As to the limit value of ck_2, observe that

$$\sqrt{A_1^2 + \chi A_2^2} - A_1 \simeq \frac{\chi A_2^2}{2A_1}$$

as $\omega \to \infty$. Now, e.g., for transverse waves, since $\mu'' \in L^1(\mathbb{R}^+)$ an integration by parts yields

$$|\omega A_2| = \left|\omega \int_0^\infty \mu'(s) \sin \omega s \, ds\right| = \left|\mu_0' + \int_0^\infty \mu''(s) \cos \omega s \, ds\right|.$$

Then by Riemann-Lebesgue's lemma we have $\omega|A_2| \to \mu_0'$, $2\mu_0' + \lambda_0'$ as $\omega \to \infty$ whence the result (6.4.17). □

Acceleration waves are characterized by a single singular surface σ and this indicates that the comparison is significant when planes of constant phase and planes of constant amplitude are parallel, namely $\chi = 1$. In such a case, (6.4.16) and (6.4.17) are equivalent to (6.1.18) and (6.1.17).

Chapter 7

Unbounded Relaxation Functions and Rayleigh Problem

7.1. Viscoelastic bodies with unbounded relaxation functions. So far we have considered relaxation functions such that $\mathbf{G}' \in L^1(\mathbb{R}^+)$ and then $\mathbf{G}_0$ is finite. Also, it is usually understood that $\mathbf{G}'_0 = \mathbf{G}'(0)$ is finite. Yet there are indications that sometimes such boundedness conditions are overly restrictive.

Indications that $\mathbf{G}'$ may have singularities arose from some rheological models motivated by molecular behaviour [114], [133], [31]. For instance, Curtiss and Bird [31] considered polymer solutions and modelled macromolecules as freely jointed bead-rod chains. Then they determined the expression for the stress tensor by taking into account the kinetic or bead momentum transport contribution, the polymer or intramolecular contribution, the polymer-polymer or intermolecular contribution. By following the approach of Irving and Kirkwood for undiluted solutions, they obtained a Boltzmann function of the form

$$G'(s) = \frac{\delta}{\tau} \sum_{n=1}^{\infty} \exp\Big[- \frac{\pi^2 (2n-1)^2}{\tau} s \Big], \tag{7.1.1}$$

τ being a characteristic time and δ a quantity proportional to the bead density. It is to be observed that the same type of relaxation function follows through the somewhat alternative approach of Doi and Edwards.

Another indication came from experimental work of Laun [90]. He started from a relation of the form

$$G'(s) = \sum_i a_i \exp\Big(- \frac{s}{\tau_i} \Big) \tag{7.1.2}$$

where the τ_i's are relaxation times. The check of (7.1.2) on LPDE melt was performed via a time-harmonic motion by accounting for eight terms only[37] through $G'_s(\omega)$ and $G'_c(\omega)$. The results turned out to be in favour of the validity of the model. Then it seems that, for certain materials, the unboundedness of G_0 and/or G'_0 has to be allowed.

In addition to these considerations of physical character, we observe that some works on the Rayleigh problem have shown that interesting, mathematical developments are involved which lead to suggestive properties of the solution. Narain and Joseph [103] have investigated the Rayleigh problem, in the one-dimensional case, in the form of step jumps in the velocity or displacement of the boundary of an incompressible fluid in a shearing motion. They have found that the discontinuity propagates into the interior with a speed $\sqrt{G_0/\rho}$ if the values G_0, G'_0 are both finite. If G_0 is finite but $G'_0 = -\infty$, then the boundary of the support of the solution still propagates with the speed $\sqrt{G_0/\rho}$. By analogy with the results of the previous chapter (cf. Theorems 6.1.3 and 6.1.4) we expect that the amplitude of the discontinuity vanishes. Really, it is found that the solutions on both sides of the boundary match together in a C^∞-fashion. Still as expected, if G_0 is finite but G'_0 vanishes, then the amplitude of the discontinuity turns out to remain constant as it does in nondissipative bodies. Finally, in the case $G_0 = \infty$, the fluid shows a parabolic behaviour in that the discontinuity is felt instantaneously throughout the fluid. Similarly analogous results have been derived by Narain and Joseph [104] in the case of viscoelastic solids.

Incidentally, by way of example and by analogy with Curtiss and Bird's expression (7.1.1), in [111] the Boltzmann function

$$G'(s) = \sum_{n=1}^{\infty} \exp(-n^\alpha s), \qquad \alpha > \tfrac{1}{2},$$

is adopted. In such a case $G'(s) \sim s^{-1/\alpha}$ as $s \to 0$ and then $G' \in L^1(\mathbb{R}^+)$ if $\alpha > 1$.

The investigations by Joseph and co-workers were usually based on a set of constitutive assumptions of the form

(J_1) $G : \mathbb{R}^+ \to \mathbb{R}^{++}$;

(J_2) G is strictly monotonically decreasing;

(J_3) $G \in C[\mathbb{R}^+) \cap PC^1(\mathbb{R}^+)$;[38]

(J_4) $G(s) = O(\exp(-\bar{\lambda} s))$ for some positive $\bar{\lambda}$ as $s \to \infty$;

(J_5) $G' < 0$ is strictly monotonically increasing and $\lim G'(s) = 0$ as $s \to \infty$.

An analysis of the Rayleigh problem for viscoelastic fluids has been developed by Renardy [111] by framing the relaxation function in a suitable

[37] With $\tau_i \in [10^{-4}\, s,\, 10^3\, s]$.

[38] P stands for piecewise.

fractional Sobolev space and letting it be nonnegative and monotone decreasing.

The conditions on monotonic behaviour seem to be much too severely restricted, though well motivated by technical reasons. Even more unduly restrictive seems to be the condition (J_4) which is usually introduced to gain convenient analytical properties of the Laplace transform. Now, the developments of Chapters 4 and 5 make us to think that even in the new situation the restrictions of thermodynamic character might be the most natural and fruitful hypotheses on the relaxation function for any investigation of the properties of the solution. Then, also with a view to analytical developments, it seems convenient to commence by deriving the thermodynamic restrictions for unbounded relaxation functions.

The adjective unbounded is used here with a generic meaning; as argued above, unboundedness may represent various situations. In this regard we mention that the theory developed by Saut and Joseph [115] provides a general setting for functionals defined through unbounded relaxation functions. The suitable "fading memory space" turns out to be a locally convex topological vector space[39] and its dual is a weighted $L^2(\mathbb{R}^+)$ along with the space of distributions, with $\{0\}$ as support, in the dual of the Sobolev space $H^j(\mathbb{R})$, for some integer j. Such setting proves to be convenient also for nonlinear functionals. Here, though, we are still interested in linear functionals only. Moreover, to make the connection with previous chapters more immediate, we keep working within the scheme of the standard fading memory space. Of course, we have to refine the specific structure of the space in order that the type of unboundedness at hand be allowed.

7.2. Modelling and thermodynamic restrictions. Consider a viscoelastic solid as described by the usual constitutive equation (3.1.2). Here we have to abandon the customary restrictions on the relaxation function $\mathbf{G}$. In particular, $\mathbf{G}'$ need not be an element of $L^1(\mathbb{R}^+)$. Accordingly, in terms of $\breve{\mathbf{G}}(s) = \mathbf{G}(s) - \mathbf{G}_\infty$, we formally integrate by parts and write

$$\mathbf{T}(t) = \mathbf{G}_\infty \mathbf{E}(t) + \int_0^\infty \breve{\mathbf{G}}(s)\, \dot{\mathbf{E}}(t-s)\, ds. \tag{7.2.1}$$

Concerning the constitutive assumption on $\mathbf{G}$, we observe that we have to allow for $\mathbf{G}_0'$ and, possibly, $\mathbf{G}_0$ unbounded. So, as a general constitutive assumption on $\mathbf{G}$, we let

$$\breve{\mathbf{G}} \in L^1(\mathbb{R}^+), \qquad \mathbf{G}_\infty \text{ bounded.} \tag{7.2.2}$$

In other words, we let $\mathbf{G}_0$, besides $\mathbf{G}_0'$, be unbounded but require that the singularity of $\breve{\mathbf{G}}(s)$ at $s = 0$ be integrable.

---

[39]Then, in particular, histories are no longer elements of a normed space.

As a domain for the functional (7.2.1), we consider a restriction of the standard fading memory space of §1.2. Specifically, letting k be the usual influence function, consider the set of histories

$$\breve{\mathcal{H}} = \left\{ \mathbf{E}^t : \mathbb{R}^+ \to \mathrm{Sym},\ |\mathbf{E}(t)|^2 + \int_0^\infty |\dot{\mathbf{E}}^t(s)|^2\, k(s)\, ds < \infty \right\}.$$

Then, on defining the norm $\|\mathbf{E}^t\|$ by

$$\|\mathbf{E}^t\|^2 = |\mathbf{E}(t)|^2 + \int_0^\infty |\dot{\mathbf{E}}^t(s)|^2\, k(s)\, ds \tag{7.2.3}$$

and the inner product $(\mathbf{E}_1^{t_1}, \mathbf{E}_2^{t_2})$ between two histories by

$$(\mathbf{E}_1^{t_1}, \mathbf{E}_2^{t_2}) = \mathbf{E}_1(t_1) \cdot \mathbf{E}_2(t_2) + \int_0^\infty \dot{\mathbf{E}}_1^{t_1}(s) \cdot \dot{\mathbf{E}}_2^{t_2}(s)\, k(s)\, ds, \tag{7.2.4}$$

the set of histories $\breve{\mathcal{H}}$ is given the structure of Hilbert space. Really $\breve{\mathcal{H}}$ becomes the Sobolev space H^1 with weight function k.

It is a straightforward exercise to ascertain that the properties (I)–(IV) in Chapter 1 of the fading memory space still hold. The space $\breve{\mathcal{H}}$ so characterized is then rightly viewed as a fading memory space.

State and process for the solid characterized by (7.2.1)–(7.2.2) are the same as those considered so far. For, the state σ is the history $\mathbf{E}^t$ and the process P is the function $\dot{\mathbf{E}}(t)$, $t \in [0, d_P)$.

We are now in a position to derive the thermodynamic restrictions on $\mathbf{G}$. By the general statement in §2.5, we consider the following requirement.

Second law for cycles. *The inequality*

$$\int_0^d \mathbf{T}(t) \cdot \dot{\mathbf{E}}(t)\, dt > 0 \tag{7.2.5}$$

holds for any nonconstant history $\mathbf{E}^t$ *such that* $\mathbf{E}^t$, $t \in [0, d)$, *is a cycle.*

Necessary conditions for the validity of the second law are given by the following theorem.

THEOREM 7.2.1. *The second law holds only if*

$$\mathbf{G}_\infty = \mathbf{G}_\infty^T, \qquad \breve{\mathbf{G}}_c(\omega) > 0, \qquad \omega \in \mathbb{R}^{++}. \tag{7.2.6}$$

Proof. Letting $\mathbf{E}_1, \mathbf{E}_2 \in \mathrm{Sym}$, consider the periodic tensor function

$$\tilde{\mathbf{E}}(t) = \mathbf{E}_1 \cos \omega t + \mathbf{E}_2 \sin \omega t, \qquad \omega \in \mathbb{R}^{++}.$$

For any finite value of ω the history $\mathbf{E}^t$ is an element of $\breve{\mathcal{H}}$. Since the period is $2\pi/\omega$, to fix ideas we let $d = 2\pi/\omega$. In view of (7.2.1), substitution in (7.2.5) yields

$$\mathbf{E}_2\cdot(\mathbf{G}_\infty - \mathbf{G}_\infty^T)\mathbf{E}_1 + \omega[\mathbf{E}_1\cdot\breve{\mathbf{G}}_c(\omega)\mathbf{E}_1 + \mathbf{E}_2\cdot\breve{\mathbf{G}}_c(\omega)\mathbf{E}_2 + \mathbf{E}_2\cdot(\breve{\mathbf{G}}_s(\omega) - \breve{\mathbf{G}}_s^T(\omega))\mathbf{E}_1] > 0.$$

The limit case $\omega \to 0$ and the arbitrariness of $\mathbf{E}_1$, $\mathbf{E}_2$ imply the symmetry of $\mathbf{G}_\infty$. If now we let $\mathbf{E}_1 = \mathbf{E}_2$, we obtain the positive definiteness of $\breve{\mathbf{G}}_c(\omega)$ for any nonzero, finite ω. □

On account of (7.2.6), henceforth we take the relaxation function $\mathbf{G}$ to satisfy

(a) $\mathbf{G}_\infty = \mathbf{G}_\infty^T > 0$;

(b) $\breve{\mathbf{G}} \in L^1(\mathbb{R}^+)$, $\breve{\mathbf{G}}_c(\omega) > 0$, $\omega \in \mathbb{R}^{++}$.

Now we show that the positive definiteness of $\breve{\mathbf{G}}_c$ forces the Laplace transform $\breve{\mathbf{G}}_L$ to be an analytic function in the complex half plane $\mathbb{C}^{++}$. To this end we need the following property.

LEMMA 7.2.1. *The Laplace transform $\breve{\mathbf{G}}_L$ and the half-range cosine transform $\breve{\mathbf{G}}_c$ are related by*

$$\breve{\mathbf{G}}_L(p) = \frac{2}{\pi}\int_0^\infty \frac{p\,\breve{\mathbf{G}}_c(\omega)}{p^2+\omega^2}\,d\omega, \qquad p \in \mathbb{C}^{++}. \tag{7.2.7}$$

Proof. Recall that if $f, h \in L^1(\mathbb{R})$ and $fh \in L^1(\mathbb{R})$ then (cf. [6], Thm. 15)

$$\int_{-\infty}^{\infty} f(u)\,h(u)\,\exp(-i\beta u)\,du = \frac{1}{2\pi}\int_{-\infty}^{\infty} f_F(\omega)\,h_F(\beta-\omega)\,d\omega \tag{7.2.8}$$

provided that the right-hand side is continuous with respect to $\beta \in \mathbb{R}$. Letting $\alpha \in \mathbb{R}^{++}$, make the identifications

$$f(u) = \breve{\mathbf{G}}(|u|), \qquad u \in \mathbb{R}, \qquad h(u) = \begin{cases} 0, & u < 0, \\ \exp(-\alpha u), & u \geq 0. \end{cases}$$

Then by repeating step by step the proof of Lemma 4.8.1, we obtain the desired result (7.2.7). □

Now we are in a position to prove the analyticity of $\breve{\mathbf{G}}_L(p)$.

THEOREM 7.2.2. *The properties $\breve{\mathbf{G}} \in L^1(\mathbb{R}^+)$ and $\breve{\mathbf{G}}_c(\omega) > 0$, $\omega \in \mathbb{R}^{++}$ imply that $\breve{\mathbf{G}}_L(p)$ is analytic on $\mathbb{C}^{++}$, is continuous on $\mathbb{C}^+$, and does not vanish on $\mathbb{C}^+ \setminus \{0\}$.*

Proof. Let $p = \alpha + i\beta$, $\alpha \in \mathbb{R}^{++}$, $\beta \in \mathbb{R}$. By (7.2.7) we can write

$$\breve{\mathbf{G}}_L(p) = \frac{2}{\pi}\int_0^\infty \frac{(\alpha+i\beta)(\alpha^2-\beta^2+\omega^2-2i\alpha\beta)}{(\alpha^2-\beta^2+\omega^2)^2+4\alpha^2\beta^2}\breve{\mathbf{G}}_c(\omega)\,d\omega.$$

Hence

$$\Re\breve{\mathbf{G}}_L(p) = \frac{2\alpha}{\pi}\int_0^\infty \frac{\alpha^2+\beta^2+\omega^2}{(\alpha^2-\beta^2+\omega^2)^2+4\alpha^2\beta^2}\breve{\mathbf{G}}_c(\omega)\,d\omega.$$

Then the positive definiteness of $\breve{\mathbf{G}}_c$ implies that $\Re\breve{\mathbf{G}}_L(p) > 0$ for any $p \in \mathbb{C}^{++}$. Furthermore, since $\breve{\mathbf{G}}_c(\beta) = \Re\breve{\mathbf{G}}_L(p)$ then, at any point $(0,\beta)$ of $\partial\mathbb{C}^+$, except $\beta = 0$, we have $\Re\breve{\mathbf{G}}_L(p) > 0$. Accordingly, $\breve{\mathbf{G}}_L(p)$ has no zeroes in the half plane $\mathbb{C}^+$, but the origin.

Finally, because $\breve{\mathbf{G}} \in L^1(\mathbb{R}^+)$, the derivative

$$\frac{d}{dp}\breve{\mathbf{G}}_L(p) = -\int_0^\infty s\exp(-p\,s)\breve{\mathbf{G}}(s)\,ds$$

exists at any $p \in \mathbb{C}^{++}$ and

$$\int_0^\infty [\exp(-p\,s) - \exp(-i\beta\,s)]\breve{\mathbf{G}}(s)\,ds \to 0$$

as $p \to i\beta$ within $\mathbb{C}^+$. So $\breve{\mathbf{G}}_L(p)$ is analytic in $\mathbb{C}^{++}$ and continuous on $\mathbb{C}^+$. □

The analyticity of $\breve{\mathbf{G}}_L(p)$ is the basic property for the developments of the next sections.

7.3. The Rayleigh problem for solids. Consider a half space described by $x \in \mathbb{R}^+$ and occupied by a viscoelastic solid as characterized in the previous section. Here we are interested in a transverse motion of the body and then, letting the body be isotropic, the only pertinent part of G will be the shear relaxation function. We write G_∞ and $\breve{G}$ for the shear part of $\mathbf{G}_\infty$ and $\breve{\mathbf{G}}$.

The Rayleigh problem for a viscoelastic solid consists in finding the solution to the transverse displacement field u on $\mathbb{R}^+ \times \mathbb{R}^+$ such that

(7.3.1)
$$\rho\frac{\partial^2 u}{\partial t^2} = G_\infty\frac{\partial^2 u}{\partial x^2} + \int_0^t \breve{G}(s)\frac{\partial^3 u}{\partial x^2\partial t}(t-s)\,ds, \qquad x \in \mathbb{R}^{++}, \qquad t \in \mathbb{R}^{++},$$

(7.3.2)
$$u(0,t) = u_0, \qquad t \in \mathbb{R}^{++},$$

(7.3.3)
$$\lim_{x\to\infty} u(x,t) = 0, \qquad t \in \mathbb{R}^{++},$$

(7.3.4)
$$u(x,0) = \frac{\partial u}{\partial t}(x,0) = 0, \qquad x \in \mathbb{R}^{++}.$$

The solution $u(x,t)$ to the problem (7.3.1)–(7.3.4) is assumed to be of bounded variation and locally integrable on $\mathbb{R}^+ \times \mathbb{R}^+$. Obviously, by (7.3.4) it follows that also $\partial u(x,0)/\partial x$ and $\partial^2 u(x,0)/\partial x^2$ vanish as $x \in \mathbb{R}^{++}$.

It is worth mentioning that in the case of smooth relaxation functions, the Rayleigh problem has long been studied; a review of the early literature on the subject is given in [16], §4.5.

Consider the Laplace transform of u,

$$u_L(x,p) := \int_0^\infty \exp(-pt)\, u(x,t)\, dt, \qquad p \in \mathbb{C}^{++},$$

where $x \in \mathbb{R}^{++}$ plays the role of a parameter. Applying the Laplace transform to (7.3.1) and on account of (7.3.4), we get

$$\rho p^2 u_L = \bar{G}(p) \frac{\partial^2 u_L}{\partial x^2}, \qquad x \in \mathbb{R}^{++}$$

where $\bar{G}(p) = G_\infty + p\, \breve{G}_L(p)$. Hence on account of (7.3.2) and (7.3.3) we have

$$u_L(x,p) = \frac{u_0}{p} \exp\left(-px\sqrt{\frac{\rho}{\bar{G}(p)}}\right). \tag{7.3.5}$$

Incidentally, at this stage replacing the $-$ sign in (7.3.5) with the $+$ sign would produce an equally acceptable solution. The reason for the choice will appear in a moment. The $-$ sign leads to a discontinuity propagating along the chosen domain $x \geq 0$ while the $+$ sign would lead to propagation along $x \leq 0$.

By Theorem 7.2.2 the Laplace transform $\breve{G}_L(p)$ is analytic in $\mathbb{C}^{++}$. Then, as $p \in \mathbb{C}^{++}$, the function (7.3.5) is analytic and we can write the sought solution $u(x,t)$ through the inverse Laplace transform

$$u(x,t) = \frac{u_0}{2\pi i} \int_{\gamma - i\infty}^{\gamma + i\infty} \frac{\exp(pt)}{p} \exp\left(- px\sqrt{\frac{\rho}{\bar{G}(p)}}\right) dp, \qquad \gamma > 0. \tag{7.3.6}$$

The evaluation of the integral in (7.3.6) parallels the procedure developed in [103] for fluids and reexamined in [104] for solids. Letting H denote the Heaviside step function, we can express the result through the following theorem.

THEOREM 7.3.1. *Because $\breve{G}_L(p)$ is analytic in $\mathbb{C}^{++}$, the inverse Laplace transform* (7.3.6) *of* (7.3.5) *is given by an expression of the form*

$$\frac{u}{u_0}(x,t) = f(x,t)\, H\left(t - \sqrt{\frac{\rho}{G_0}}x\right). \tag{7.3.7}$$

Proof. Since $\breve{G}$ is analytic then $\bar{G}$ is analytic, too; the analyticity of $\bar{G}$ is at the basis of the proof.

Let $\nu = \sqrt{\rho/G_0}$. As $t - \nu x < 0$, we can choose the contour comprising the segment between $\gamma - iR$ and $\gamma + iR$ and the semicircular path Γ_R, of radius R, centred at $\gamma + i0$ with $\Re p \geq \gamma$. We observe that, on Γ_R,

$$p\breve{G}_L(p) = \breve{G}(0) + \varepsilon(R)$$

where ε is a complex-valued quantity and $|\varepsilon(R)| \to 0$ as $R \to \infty$. Then we have

$$\int_{\Gamma_R} \frac{1}{p} \exp\left[p\left(t - x\sqrt{\rho/\bar{G}(p)}\right)\right] dp = \int_{\Gamma_R} \frac{1}{p} \exp\left[p(t - x\sqrt{\rho/[G_0 + \varepsilon(R)]})\right] dp.$$

Letting $-\tau = t - \nu x < 0$, we have the estimate

$$\left|\int_{\Gamma_R} \frac{1}{p} \exp\left[p\left(t - x\sqrt{\rho/\bar{G}(p)}\right)\right] dp\right| \leq \int_{-\pi/2}^{\pi/2} \frac{1}{R} \exp[(-\tau + \delta(R))(\gamma + R\cos\phi)] d\phi$$

where $\delta(R) \to 0$ as $R \to \infty$. Then

$$\int_{\Gamma_R} \frac{1}{p} \exp\left[p\left(t - x\sqrt{\rho/\bar{G}(p)}\right)\right] dp \to 0 \quad \text{as } R \to \infty.$$

The application of Cauchy's theorem to the contour shows that the integral $\int_{\gamma-iR}^{\gamma+iR} + \int_{\Gamma_R}$ of $\exp[p(t - x\sqrt{\rho/\bar{G}(p)})]$ vanishes. The limit $R \to \infty$ provides the result

$$u(x,t) = 0 \quad \text{as } t - \nu x < 0. \tag{7.3.8}$$

As $t - \nu x > 0$ we choose as contour the rectangle with vertices $0 - iR, \gamma - iR, \gamma + iR, 0 + iR$, and the segment between $0 - ir$ and $0 + ir$, $r < R$, replaced by the semicircular path Γ_r, centered at the origin with radius r, in $\mathbb{C}^+$. Now, denote by γ_+ the segment between $\gamma + iR$ and $0 + iR$. As $p \in \gamma_+$ we have $p = \alpha + i\gamma$ and then

$$\left|\int_{\gamma_+} \frac{1}{p} \exp\left[p\left(t - x\sqrt{\rho/\bar{G}(p)}\right)\right] dp\right| \leq \int_0^{\gamma} \frac{1}{\sqrt{\alpha^2 + R^2}} \exp[\alpha(t - x\sqrt{\rho/[G_0 + \varepsilon(R)]})] d\alpha.$$

Then the integral over γ_+ tends to zero as R tends to infinity. The same is true for the integral over the symmetric segment γ_- between $0 - iR$ and $\gamma - iR$. To evaluate the integral over the semicircular path Γ_r, taken clockwise, let $p = r\exp(i\theta)$, $\theta \in [\frac{\pi}{2}, -\frac{\pi}{2}]$. Then, as $r \to 0$,

$$\lim_{r\to 0} \frac{1}{2\pi i} \int_{\Gamma_r} \frac{1}{p} \exp\left[p\left(t - x(\sqrt{\rho/\bar{G}(p)}\right)\right] dp = \frac{1}{2\pi} \int_{\pi/2}^{-\pi/2} [1 + O(\sqrt{\delta})] d\theta = -\frac{1}{2}.$$

The application of Cauchy's theorem to the contour and account of (7.3.6) gives

$$\frac{u}{u_0}(x,t) = \frac{1}{2} + \frac{1}{2\pi i}\int_{-\infty}^{\infty} \frac{1}{\beta} \exp\left[i\beta\left(t - x\sqrt{\rho/\bar{G}(i\beta)}\right)\right] d\beta. \tag{7.3.9}$$

By (7.2.8) we have $\bar{G}(i\beta) = G_\infty + \beta\breve{G}_s(\beta) + i\beta\breve{G}_c(\beta)$ so that $\Re\bar{G}(i\beta)$ is even in β while $\Im\bar{G}(i\beta)$ is odd. Accordingly, letting

$$\bar{G}(i\beta) = q(\beta)\exp(ia(\beta))$$

where $q(\beta) = |\bar{G}(i\beta)|$ and $a(\beta) = \arg\bar{G}(i\beta)$, we see that $q(\beta)$ is even in β and $a(\beta)$ is odd and, moreover, $a(\beta) \in (0,\pi)$ as $\beta \in \mathbb{R}^{++}$. Now observe that

$$\exp[-i\beta x\sqrt{\rho/q(\beta)\exp(ia(\beta))}] = \begin{cases} \exp\{-\beta x\sqrt{\rho/q(\beta)}\exp[\tfrac{1}{2}i(\pi - a(\beta))]\}, & \beta > 0, \\ \exp\{-|\beta| x\sqrt{\rho/q(|\beta|)}\exp[-\tfrac{1}{2}i(\pi - a(|\beta|))]\}, & \beta < 0. \end{cases}$$

Then, upon some rearrangement, we obtain

$$\begin{aligned} &\frac{1}{2\pi i}\int_{-\infty}^{\infty} \frac{1}{\beta}\exp\left[i\beta\left(t - x\sqrt{\rho/\bar{G}(i\beta)}\right)\right] d\beta \\ &\quad = \frac{1}{\pi}\int_0^\infty \frac{1}{\beta}\exp[-\beta x\sqrt{\rho/q(\beta)}\cos\tfrac{1}{2}(\pi - a(\beta))] \\ &\qquad\qquad \sin[\beta(t - x\sqrt{\rho/q(\beta)})\sin\tfrac{1}{2}(\pi - a(\beta))]d\beta. \end{aligned}$$

Accordingly, (7.3.8) and (7.3.9) are expressed in the form (7.3.7) where

$$\begin{aligned} f(x,t) = \frac{1}{2} + \frac{1}{\pi}\int_0^\infty &\frac{1}{\beta}\exp[-\beta x\sqrt{\rho/q(\beta)}\cos\tfrac{1}{2}(\pi - a(\beta))] \\ &\sin[\beta(t - x\sqrt{\rho/q(\beta)})\sin\tfrac{1}{2}(\pi - a(\beta))]d\beta, \end{aligned}$$

as $t > \nu x$. □

This means that a discontinuity can occur at x at time $t = \nu x$. Indeed, the discontinuity is determined by evaluating $f(x, \nu x^+)$, where x^+ is a reminder that the limit is performed as $t \to \nu x$ while $t > \nu x$. It is then of interest to determine $f(x, \nu x^+)$.

THEOREM 7.3.2. *The discontinuity $f(x, \nu x^+)$ is given by*

$$f(x, \nu x^+) = \exp\left(\frac{\nu x G_0'}{2G_0}\right). \tag{7.3.10}$$

Proof. By applying Watson's lemma in a particular way (cf. [113], §4.4 with $\lambda = 0$) we can say that if

$$\breve{G}(s) = \breve{G}_0 + \breve{G}_0' s + O(s^2)$$

as $s \in \mathbb{R}^+$ is in a neighbourhood of the origin $s = 0$, then $p\breve{G}_L(p)$ has the asymptotic expansion

$$p\breve{G}_L(p) = \breve{G}_0 + \frac{\breve{G}_0'}{p} + O\Big(\frac{1}{p^2}\Big),$$

for $|\arg p| < \frac{\pi}{2} - \delta$, $\delta > 0$. Accordingly we evaluate $u(x,t)$ through the integral along the line $\gamma - i\infty$, $\gamma + i\infty$, with $\gamma > 0$. Since

$$G_\infty + \breve{G}_0 = G_0, \qquad \breve{G}_0' = G_0',$$

and

$$\sqrt{\Big(\rho/G_0 + \frac{G_0'}{p} + O\Big(\frac{1}{p^2}|\Big)\Big)} = \nu\Big[1 - \frac{1}{2}\frac{G_0'}{G_0}\frac{1}{p} + O\Big(\frac{1}{p^2}\Big)\Big],$$

we obtain from (7.3.6) that

$$\begin{aligned}\frac{u}{u_0}(x,t) &= \exp\left[\frac{\nu x G_0'}{2G_0}\right]\frac{1}{2\pi i}\int_{\gamma - i\infty}^{\gamma + i\infty}\frac{1}{p}\exp[p(t-\nu x)]\Big[1 + O\Big(\frac{1}{p}\Big)\Big]\,dp \\ &= \exp\left[\frac{\nu x G_0'}{2G_0}\right] H(t - \nu x) \\ &\quad + \frac{1}{2\pi i}\int_{\gamma - i\infty}^{\gamma + i\infty}\exp\left[\frac{\nu x G_0'}{2G_0}\right]\frac{1}{p}\exp[p(t-\nu x)]\,O\Big(\frac{1}{p}\Big)\,dp.\end{aligned}$$

The last integral is continuous in $x, t \in \mathbb{R}^+$. Then (7.3.10) follows. □

It is worth commenting on the results (7.3.7) and (7.3.10). If both $G_0 \in \mathbb{R}^{++}$ and $G_0' \in \mathbb{R}^-$ are finite, then the jump in u induced by the boundary data decays while propagating with the constant speed $1/\nu$. If, in particular, $G_0' = 0$ then the jump is constant. By (7.3.10), at the limit case when G_0 is finite but $G_0' = -\infty$, the jump vanishes and the solution is smooth. Yet by (7.3.7) the boundary of the support of u still propagates with the speed $1/\nu$. Finally, at the limit case $G_0 = \infty$, the solution is still smooth in that the initial jump is instantly smoothed.

Incidentally, as one might have expected, the result (7.3.10) for the discontinuity in the Rayleigh problem is just the one-dimensional counterpart of the general expression (6.1.11), or the specific relations (6.1.17) and (6.1.18), for the amplitude of discontinuity waves.

We append a comment on the behaviour of the derivative $\partial f/\partial t$ at the discontinuity. By Watson's lemma, if

$$\breve{G}(s) = \breve{G}_0 + \breve{G}_0' s + \cdots + \frac{1}{n!}\breve{G}_0^{(n)} s^n + O(s^{n+1}),$$

then

$$p\,\breve{G}_L(p) = \breve{G}_0 + \frac{\breve{G}_0'}{p} + \cdots + \frac{\breve{G}_0^{(n)}}{p^n} + O\Big(\frac{1}{p^{n+1}}\Big).$$

Then a generalization of the previous procedure yields the derivatives of f at $(x, \nu x^+)$ up to order $n-1$ in terms of the derivatives up to order n of $G(s)$ at $s = 0$. For instance (see [103]),

$$\frac{\partial f}{\partial t}(x, \nu x^+) = -\nu x \exp\left(\frac{\nu x G_0'}{G_0}\right)\left[\frac{3}{8}\left(\frac{G_0'}{G_0}\right)^2 - \frac{1}{2}\frac{G_0''}{G_0}\right]$$

where $G_0'' = G''(0)$. The same result, in the forms (6.2.17) and (6.2.20), has been found by applying the compatibility relations for induced discontinuities.

7.4. The Rayleigh problem for fluids. The constitutive equation (7.2.1) may serve formally for viscoelastic fluids as well simply by letting $\mathbf{G}_\infty = 0$, replacing $\dot{\mathbf{E}}$ with $\mathbf{D}$, and adding the pressure term $-\Pi\,\mathbf{1}$. We know already that the thermodynamic restrictions are summarized by $\mathbf{G}_c(\omega) > 0$, $\omega \in \mathbb{R}^{++}$. So formally we take the condition (b) as for solids (but now $\mathbf{G} = \breve{\mathbf{G}}$) and replace (a) with $\mathbf{G}_\infty = 0$, namely

(a$\prime$) $\mathbf{G} \in L^1(\mathbb{R}^+)$, $\mathbf{G}_\infty = 0$, $\mathbf{G}_c(\omega) > 0$, $\omega \in \mathbb{R}^{++}$.

Let the fluid occupy again the half space $x \in \mathbb{R}^+$ and let the pertinent functions depend on the space variables through x only. The fluid is regarded as compressible but we confine to the linear approximation (cf. §5.10). Accordingly, upon integration over $[0, t]$, the continuity equation provides

$$\rho(x,t) - \rho(x,0) = -\rho_0 \int_0^t \nabla\cdot\mathbf{v}(x,\tau)\,d\tau, \qquad x \in \mathbb{R}^{++},$$

or, in terms of the displacement $\mathbf{u}$, $\mathbf{v} = \dot{\mathbf{u}}$,

$$\rho(x,t) - \rho(x,0) = -\rho_0[\nabla\cdot\mathbf{u}(x,t) - \nabla\cdot\mathbf{u}(x,0)], \qquad x \in \mathbb{R}^{++}.$$

For such a fluid we investigate the piston problem which consists of finding the solution to the longitudinal displacement field u on $\mathbb{R}^+ \times \mathbb{R}^+$ such that

(7.4.1)

$$\rho_0\frac{\partial^2 u}{\partial t^2} = \rho_0 c^2 \frac{\partial^2 u}{\partial x^2} + \int_0^t G(s)\frac{\partial^3 u}{\partial x^2 \partial t}(t-s)\,ds, \qquad x \in \mathbb{R}^{++}, \qquad t \in \mathbb{R}^{++},$$

where $c^2 = \Pi_\rho$, along with the initial-boundary conditions (7.3.2)–(7.3.4). Application of the Laplace transform to (7.4.1) yields

$$\frac{\partial^2 u_L}{\partial x^2} = \frac{\rho_0 p^2}{\rho_0 c^2 + p\,G_L(p)} u_L, \qquad x \in \mathbb{R}^{++}.$$

Hence, letting $\rho_0 \to \rho$ and $\bar{G}(p) = \rho_0 c^2 + p\,G_L(p)$ and accounting for (7.3.2), we find formally (7.3.5). So we can restate word by word the content of the previous section and rewrite the corresponding results. In particular, we have

$$\nu = \sqrt{\frac{\rho_0}{\rho_0 c^2 + G_0}} \tag{7.4.2}$$

and

$$f(x, \nu x^+) = \exp\left(\frac{\nu\, x\, G_0'}{2(\rho_0 c^2 + G_0)}\right). \tag{7.4.3}$$

According to (7.4.2), the jump in u propagates with a speed which shows the correction, due to the memory functional, on the value for the elastic fluid.

Formally the Rayleigh problem may be viewed as a particular case of the piston problem. As with §7.3, the Rayleigh problem involves a solenoidal motion. In fact, the equations for the Rayleigh problem are just those for the piston problem but with u denoting the transverse displacement and $c = 0$. That is why we omit the analysis of the Rayleigh problem for compressible fluids.

With regard to the Rayleigh problem for incompressible fluids, we are content with mentioning the detailed investigation by Narain and Joseph [103]. The pertinent results follow formally from those for solids by letting $G_\infty = 0$ and are essentially the same as those for compressible fluids.

We conclude this section by mentioning some recent results. Hrusa and Renardy [82] have considered the evolution of step-jump in the initial data in a linear viscoelastic material. They have shown that the propagating discontinuity is smoothed in a manner which depends on the nature of the singularity of the relaxation function, namely of G'. In the limiting case that G' has a logarithmic singularity, the solution gains smoothness gradually as time increases. More recently, Desch and Grimmer [42] have provided a general theorem characterizing the kernels that lead to smoothing of the solution of a first-order integrodifferential equation whose purely differential part is of hyperbolic type. Moreover, they have found a formula for the space interval needed to gain one degree of differentiability in the limiting case. Prüss [108] has considered the Rayleigh problem in connection with a class of relaxation functions having integrable singularities. Indeed, the second spatial derivative is replaced by a generator of a cosine family while the relaxation function is replaced by a specific Stieltjes measure that includes the case of completely monotone functions. In essence, he has proved that $v(x,\cdot)$ is of bounded variation and has a limit, which is shown to be zero, as $t \to \nu x^+$.

In these works, the assumptions are quite technical. Perhaps the adoption of hypotheses somewhat related to thermodynamic properties might lead to simpler procedures and more immediate results.

Appendix

Précis of the Properties of the Relaxation Function

For the benefit of the reader desiring a quick reference, a self-contained, concise account is exhibited of the properties of the relaxation function of a linear viscoelastic solid. Some results are restated; others are given here for the first time. Meanwhile proofs are given for results which have already been applied in this note without proof. This account stems from a research performed by Giorgi [63].

Start from the constitutive equation

$$\mathbf{T}(t) = \mathbf{G}_0\mathbf{E}(t) + \int_0^\infty \mathbf{G}'(s)\,\mathbf{E}(t-s)\,ds$$

where $\mathbf{G}_0$ and $\mathbf{G}'(s)$, $s \in \mathbb{R}^+$, are elements of Lin(Sym). To let constant histories correspond to finite values of the stress $\mathbf{T}$, we assume $\mathbf{G}' \in L^1(\mathbb{R}^+)$. The relaxation function

$$\mathbf{G}(s) = \mathbf{G}_0 + \int_0^s \mathbf{G}'(u)\,du$$

can also be written as

$$\mathbf{G}(s) = \mathbf{G}_\infty - \int_s^\infty \mathbf{G}'(u)\,du.$$

We let

$$\mathbf{G}_\infty > 0$$

for solids. Moreover, we let

$$\mathbf{G}' = \mathbf{G}'^T;$$

the case when the symmetry assumption is not made will also be considered.

The Fourier transform $\mathbf{G}'_F(\omega)$ can be written as

$$\mathbf{G}'_F(\omega) = \int_0^\infty \mathbf{G}'(u)\exp(-i\omega u)\,du = \mathbf{G}'_c(\omega) - i\mathbf{G}'_s(\omega).$$

Concerning the Laplace transform, throughout we consider the complex variable p in the form $p = \alpha + i\beta$, $\alpha \in \mathbb{R}^+$ (or $\mathbb{R}^{++}$), $\beta \in \mathbb{R}$. We have

$$\mathbf{G}'_L(p) = \tilde{\mathbf{G}}'_c(\alpha,\beta) - i\tilde{\mathbf{G}}'_s(\alpha,\beta)$$

where

$$\tilde{\mathbf{G}}'_c(\alpha,\beta) = \int_0^\infty \mathbf{G}'(u)\exp(-\alpha u)\,\cos\beta u\,du,$$

$$\tilde{\mathbf{G}}'_s(\alpha,\beta) = \int_0^\infty \mathbf{G}'(u)\exp(-\alpha u)\,\sin\beta u\,du.$$

LEMMA A.1. *The functions $\mathbf{G}'_c, \mathbf{G}'_s$ are continuous and bounded on $\mathbb{R}$. Further, if $\mathbf{G}'$ is absolutely continuous, namely $\mathbf{G}'' \in L^1(\mathbb{R})$, then*

$$\text{(A.1)} \qquad \lim_{\omega\to\infty} \omega\mathbf{G}'_s(\omega) = \mathbf{G}'(0^+), \qquad \lim_{\omega\to\infty} \omega\mathbf{G}'_c(\omega) = 0$$

whereas, if $\int_0^\infty s|\mathbf{G}'(s)|ds < \infty$ then

$$\text{(A.2)} \qquad \mathbf{G}'_s(\omega) = O(\omega) \quad \text{as} \quad \omega \to 0.$$

Proof. Since $\mathbf{G}' \in L^1(\mathbb{R})$ then classical theorems on the Fourier transform imply the continuity and boundedness of $\mathbf{G}'_F$ on $\mathbb{R}$. The continuity and boundedness of $\mathbf{G}'_c$ and $\mathbf{G}'_s$ are a trivial consequence.

If $\mathbf{G}'' \in L^1(\mathbb{R}^+)$, an integration by parts gives

$$\int_0^\infty \mathbf{G}''(s)\sin\omega s\,ds = -\mathbf{G}'(s)\frac{\cos\omega s}{\omega}\Big|_0^\infty + \frac{1}{\omega}\int_0^\infty \mathbf{G}''(s)\cos\omega s\,ds.$$

Hence we have

$$\omega\mathbf{G}'_s(\omega) = \mathbf{G}'(0^+) + \mathbf{G}''_c(\omega)$$

whence $(\text{A.1})_1$. By the same token we have

$$\mathbf{G}'_c(\omega) = -\frac{\mathbf{G}''_s(\omega)}{\omega}$$

and then $(A.1)_2$.

Finally, look at $\mathbf{G}'_s$ in the form

$$\mathbf{G}'_s(\omega) = \omega \int_0^\infty u\ \mathbf{G}'(u)\frac{\sin \omega u}{\omega\ u}\, du.$$

Since $\mathbf{G}' \in L^1(\mathbb{R}^+)$ and $|\sin \omega u/\omega u| \leq 1$ for any $u, \omega \in \mathbb{R}$, it follows that (A.2) holds. □

For any $u \in \mathbb{R}^+$ consider $\breve{\mathbf{G}}(u) = \mathbf{G}(u) - \mathbf{G}_\infty$ and then

$$\breve{\mathbf{G}}_c(\omega) = \int_0^\infty [\mathbf{G}(u) - \mathbf{G}_\infty] \cos \omega u\, du.$$

LEMMA A.2. *The function* $\breve{\mathbf{G}}_c$ *is continuous and absolutely integrable on* $\mathbb{R}^+$. *Moreover, for any* $u \in \mathbb{R}^+$,

$$\text{(A.3)} \qquad \mathbf{G}(u) = \mathbf{G}_\infty + \frac{2}{\pi}\int_0^\infty \breve{\mathbf{G}}_c(\omega) \cos \omega u\, d\omega.$$

Proof. An integration by parts gives

$$\int_0^\infty \mathbf{G}'(u) \sin \omega u\, du = -\omega \int_0^\infty [\mathbf{G}(u) - \mathbf{G}_\infty] \cos \omega u\, du$$

whence

$$\text{(A.4)} \qquad \breve{\mathbf{G}}_c(\omega) = -\frac{1}{\omega}\mathbf{G}'_s(\omega).$$

Then, by Lemma A.1, $\breve{\mathbf{G}}_c$ is continuous and bounded in $(\bar{\omega}, \infty)$, $\bar{\omega} > 0$, and integrable on $\mathbb{R}^+$ in that

$$\lim_{|\omega| \to \infty} \omega \breve{\mathbf{G}}_c(\omega) = \lim_{\omega \to 0} \omega \breve{\mathbf{G}}_c(\omega) = 0.$$

Then (cf. [6], Chap. III), for any $u \in \mathbb{R}^+$, the inversion formula (A.3) holds.

LEMMA A.3. (Potential Transform). *For any complex* $p \in \mathbb{C}^{++}$, $\mathbf{G}'_L(p)$ *is related to* $\mathbf{G}'_s$ *and* $\mathbf{G}'_c$ by

$$\text{(A.5)} \qquad \mathbf{G}'_L(p) = \frac{2}{\pi}\int_0^\infty \frac{\omega\, \mathbf{G}'_s(\omega)}{p^2 + \omega^2}\, d\omega,$$

$$\text{(A.6)} \qquad \mathbf{G}'_L(p) = \frac{2}{\pi}\int_0^\infty \frac{p\, \mathbf{G}'_c(\omega)}{p^2 + \omega^2}\, d\omega.$$

Proof. First recall that (cf. [6], Thm. 15), if $f, h \in L^1(\mathbb{R})$ and $fh \in L^1(\mathbb{R})$ then

$$\text{(A.7)} \qquad \int_{-\infty}^{\infty} f(u)h(u)\exp(-i\beta u)\,du = \frac{1}{2\pi}\int_{-\infty}^{\infty} f_F(\omega)h_F(\beta-\omega)\,d\omega$$

provided that the right-hand side is continuous with respect to $\beta \in \mathbb{R}$. Now, letting $\alpha \in \mathbb{R}^{++}$, make the identifications

$$f(u) = \begin{cases} -\mathbf{G}'(-u), & u < 0, \\ \mathbf{G}'(u), & u \geq 0, \end{cases} \qquad h(u) = \begin{cases} 0, & u < 0, \\ \exp(-\alpha u), & u \geq 0. \end{cases}$$

Both f and h belong to $L^1(\mathbb{R})$. Moreover,

$$h_F(\omega) = \frac{1}{\alpha + i\omega}$$

so that

$$f_F(\omega)h_F(\beta-\omega) = \mathbf{G}'_F(\omega)\frac{1}{\alpha + i(\beta-\omega)}$$

and the right-hand side is apparently continuous with respect to β. Then (A.7) is applicable and we have

$$\int_0^{\infty} \mathbf{G}'(u)\exp[-(\alpha+i\beta)u]du = \frac{1}{2\pi}\int_{-\infty}^{\infty} \frac{\mathbf{G}'_F(\omega)}{p - i\omega}\,d\omega$$

whence

$$\mathbf{G}'_L(p) = -\frac{i}{\pi}\int_{-\infty}^{\infty} \frac{\mathbf{G}'_s(\omega)}{p-i\omega}\,d\omega, \qquad \Re p > 0.$$

Since

$$\frac{\mathbf{G}'_s(\omega)}{p-i\omega} = \frac{p}{p^2+\omega^2}\mathbf{G}'_s(\omega) + i\frac{\omega}{p^2+\omega^2}\mathbf{G}'_s(\omega),$$

the observation that $\mathbf{G}'_s(\omega)$ is odd in ω provides the result (A.5).

By the same token, and by analogy with Lemma 7.2.1, the identification $f(u) = \mathbf{G}'(|u|)$, $u \in \mathbb{R}$, leads to (A.6). □

Remark A.1. If $f, g \in L^1(\mathbb{R}) \cap L^2(\mathbb{R})$ then, by Plancherel's theorem,

$$\int_{-\infty}^{\infty} f(u)\,g^*(u)\,du = \frac{1}{2\pi}\int_{-\infty}^{\infty} f_F(\omega)\,g_F^*(\omega)\,d\omega$$

where $*$ denotes the complex conjugate. Letting $g(u) = h(u)\exp(i\beta u)$ with $\beta \in \mathbb{R}$ and $h : \mathbb{R}^+ \to \mathbb{R}$ we obtain (A.6). So, Lemma A.3 may be viewed as a weaker version of what can be obtained by applying Plancherel's theorem.

On account of the relations (A.4) and (A.5), we have the following result.

COROLLARY A.1. *For any* $p \in \mathbb{C}^{++}$, $\mathbf{G}'_L$ *is given by* $\breve{\mathbf{G}}_c$ *as*

$$\mathbf{G}'_L(p) = -\frac{2}{\pi}\int_0^\infty \frac{\omega^2 \breve{\mathbf{G}}_c(\omega)}{p^2+\omega^2}d\omega.$$

With these preliminaries about the Fourier and the Laplace transforms, we can derive detailed relations from the thermodynamic conditions, namely[40]

$$\mathbf{G}_0 = \mathbf{G}_0^T, \qquad \mathbf{G}_\infty = \mathbf{G}_\infty^T, \tag{A.8}$$

$$\omega \mathbf{G}'_s(\omega) \le 0 \quad \text{as } \omega \in \mathbb{R}, \tag{A.9}$$

along with the constitutive property $\mathbf{G}_\infty > 0$. Observe that (A.9) is just the weak condition (3.4.1) rather than (3.2.11).

THEOREM A.1. *If* $\mathbf{G}'$ *satisfies* (A.9) *then*

$$\beta \mathbf{G}'_s(\alpha,\beta) < 0, \qquad \beta \in \mathbb{R}\setminus\{0\}, \qquad \alpha \in \mathbb{R}^{++}. \tag{A.10}$$

Proof. We know that

$$\tilde{\mathbf{G}}'_s(\alpha,\beta) = -\Im \mathbf{G}'_L(p).$$

Then by (A.5) we have

$$\tilde{\mathbf{G}}'_s(\alpha,\beta) = \frac{4\alpha\beta}{\pi}\int_0^\infty \frac{\omega \mathbf{G}'_s(\omega)}{(\alpha^2-\beta^2+\omega^2)^2+4\alpha^2\beta^2}d\omega. \tag{A.11}$$

The right-hand side cannot vanish; otherwise $\mathbf{G}'$ should vanish almost everywhere on $\mathbb{R}^+$. Hence we have the result (A.10). □

THEOREM A.2. *If* $\mathbf{G}'$ *satisfies* (A.9) *then*

$$\mathbf{G}_\infty - \mathbf{G}_0 < \tilde{\mathbf{G}}'_c(\alpha,0) < 0. \tag{A.12}$$

Proof. Since $\Re \mathbf{G}'_L(p) = \tilde{\mathbf{G}}'_c(\alpha,\beta)$, by (A.5) we obtain

$$\tilde{\mathbf{G}}'_c(\alpha,\beta) = \frac{2}{\pi}\int_0^\infty \frac{(\alpha^2-\beta^2+\omega^2)\omega \mathbf{G}'_s(\omega)}{(\alpha^2-\beta^2+\omega^2)^2+4\alpha^2\beta^2}d\omega. \tag{A.13}$$

Hence by (A.10) we have $\tilde{\mathbf{G}}'_c(\alpha,0) < 0$. Now, by definition,

$$\tilde{\mathbf{G}}'_c(\alpha,0) = \int_0^\infty \mathbf{G}'(u)\exp(-\alpha u)du$$

[40] We recall that the symmetry of $\mathbf{G}'$ has been assumed at the outset.

and then an integration by parts yields

$$\tilde{\mathbf{G}}_c'(\alpha,0)=\alpha\int_0^\infty(\mathbf{G}(u)-\mathbf{G}_\infty)\exp(-\alpha u)du\ +\ \mathbf{G}_\infty-\mathbf{G}_0.$$

By (A.3) we have

$$\tilde{\mathbf{G}}_c'(\alpha,0)=\frac{2\alpha}{\pi}\int_0^\infty\int_0^\infty\breve{\mathbf{G}}_c(\omega)\exp(-\alpha u)\cos\omega u\ d\omega\,du\ +\ \mathbf{G}_\infty-\mathbf{G}_0.$$

By exchanging the order of integration[41] we obtain

$$\tilde{\mathbf{G}}_c'(\alpha,0)=\frac{2\alpha^2}{\pi}\int_0^\infty\frac{\breve{\mathbf{G}}_c(\omega)}{\alpha^2+\omega^2}\,d\omega+\mathbf{G}_\infty-\mathbf{G}_0.$$

By (A.4) and (A.9) it follows $\tilde{\mathbf{G}}_c'(\alpha,0)>\mathbf{G}_\infty-\mathbf{G}_0$ and the theorem is proved. □

For formal convenience we extend the domain of definition of the relaxation function $\mathbf{G}$ to $\mathbb{R}$ by letting $\mathbf{G}$ vanish on $\mathbb{R}^{--}$.

With a slight abuse of notation, we denote the new function by the same symbol $\mathbf{G}$. So $\mathbf{G}$ is a function of bounded variation and we define the Laplace-Stieltjes transform $\mathbf{G}_{LS}$ of $\mathbf{G}$ as

$$\mathbf{G}_{LS}(p)=\int_{-\infty}^\infty\exp(-pu)\,d\mathbf{G}(u)$$

whence

$$\mathbf{G}_{LS}(p)=\mathbf{G}_0+\mathbf{G}_L'(p).\tag{A.14}$$

THEOREM A.3. *If* $\mathbf{G}$ *satisfies* (A.9) (*and hence* (A.10)), *then*

$$\mathbf{G}(u)<\mathbf{G}_0,\qquad u\in\mathbb{R}^{++},\tag{A.15}$$

$$|\mathbf{E}\cdot\mathbf{G}_{LS}(p)\mathbf{E}|>0,\qquad \mathbf{E}\in\mathrm{Sym}\setminus\{\mathbf{0}\},\qquad p\in\mathbb{C}^{++},\tag{A.16}$$

$$\Re(p^*\mathbf{G}_{LS}(p))>0,\qquad p\in\mathbb{C}^{++}.\tag{A.17}$$

Proof. It follows from (A.3) that

$$\mathbf{G}_0=\mathbf{G}_\infty+\frac{2}{\pi}\int_0^\infty\breve{\mathbf{G}}_c(\omega)\,d\omega.\tag{A.18}$$

[41]Which we can because, by Lemma A.2, $\breve{\mathbf{G}}_c(\omega)$ is absolutely integrable.

Then, on using again (A.3), for any $u \in \mathbb{R}^+$ we have

$$\mathbf{G}_0 - \mathbf{G}(u) = \frac{2}{\pi}\int_0^\infty \breve{\mathbf{G}}_c(\omega)(1-\cos\omega u)\,d\omega.$$

Owing to (A.4) and (A.9), we obtain (A.15), i.e., (3.2.12).

It follows from (A.14) that $\mathbf{G}_{LS}(p)$ is analytic in the half plane $\mathbb{C}^{++}$ and continuous on $\mathbb{C}^+$. Now, in view of (A.12),

$$\mathbf{G}_{LS}(p) = \mathbf{G}_0 + \tilde{\mathbf{G}}_c'(\alpha,\beta) - i\tilde{\mathbf{G}}_s'(\alpha,\beta). \tag{A.19}$$

Then, for any $\alpha \in \mathbb{R}^{++}$, by (A.10) we have

$$\beta \in \mathbb{R}^{++} \quad \Longrightarrow \quad \tilde{\mathbf{G}}_s'(\alpha,\beta) < 0, \qquad \Im\mathbf{G}_{LS}(p) > 0,$$

$$\beta \in \mathbb{R}^{++} \quad \Longrightarrow \mathbf{G}_{LS}(p) = \mathbf{G}_0 + \tilde{\mathbf{G}}_c'(\alpha,0), \qquad \Re\mathbf{G}_{LS}(p) > 0.$$

Hence we have (A.16).

Finally, by (A.12) and (A.19) we can write

$$p^*\mathbf{G}_{LS}(p) = \alpha[\mathbf{G}_0 + \tilde{\mathbf{G}}_c'(\alpha,\beta)] - \beta b\tilde{G}_s'(\alpha,\beta).$$

In view of (A.11) and (A.13), we obtain

$$\Re[p^*\mathbf{G}_{LS}(p)] = \alpha\left\{\mathbf{G}_0 + \frac{2}{\pi}\int_0^\infty \left[1 - |p|^2\frac{\omega^2+|p|^2}{|\omega^2+p^2|^2}\right]\frac{\mathbf{G}_s'(\omega)}{\omega}\,d\omega\right\}.$$

Then by (A.18) and (A.4) we have

$$\Re[p^*\mathbf{G}_{LS}(p)] = \alpha\left[\mathbf{G}_\infty + \frac{2}{\pi}|p|^2\int_0^\infty \frac{\omega^2+|p|^2}{|\omega^2+p^2|^2}\breve{\mathbf{G}}_c(\omega)\,d\omega\right].$$

By (A.4) and the hypothesis (A.9) we obtain the result (A.17). □

Remark A.2. As an aside, consider the case when the extra condition of symmetry of $\mathbf{G}'$ is not assumed. In such a case, the thermodynamic requirements consist of the symmetry conditions (A.8) along with the inequality

$$\begin{gathered}\mathbf{E}_1\cdot\mathbf{G}_s'(\omega)\mathbf{E}_1 + \mathbf{E}_2\cdot\mathbf{G}_s'(\omega)\mathbf{E}_2 + \mathbf{E}_1\cdot[\mathbf{G}_c'(\omega) - \mathbf{G}_c'^T(\omega)]\mathbf{E}_2 \le 0,\\ \mathbf{E}_1,\,\mathbf{E}_2 \in \mathrm{Sym}, \qquad \omega \in \mathbb{R}^{++},\end{gathered} \tag{A.20}$$

which replaces (A.9). Yet the particular choice $\mathbf{E}_1 = \mathbf{E}_2 = \mathbf{E}$ and the observation that $\mathbf{G}_s'(\omega)$ is odd yields (A.9) in Sym. So it would seem that, although $\mathbf{G}'$ is not symmetric, we obtain the same consequences as though $\mathbf{G}'$ were symmetric. However, it is apparent that (A.15) no longer holds if $\mathbf{G}'$ is not symmetric.

Letting again $\mathbf{G}'$ be symmetric, we conclude this account with some properties of $\mathbf{G}_{LS}$. The Laplace transform of the constitutive equation for $\mathbf{T}$ gives

$$\mathbf{T}_L(p) = \mathbf{G}_{LS}(p)\mathbf{E}_L(p)$$

and, meanwhile, the equation of motion with given initial data takes the form (cf. (4.6.5))

$$p^2\mathbf{u}_L(p) - \nabla \cdot \mathbf{G}_{LS}(p)\nabla\mathbf{u}_L(p) = \mathbf{B}_L(p).$$

By (A.16),[42] the operator in the left-hand side is strongly elliptic in $H_0^1(\Omega)$ for every $p \in \mathbb{C}^{++}$ and this property is essential in the investigation of the asymptotic behaviour of the solution (cf. §4.6). A more restrictive condition, namely

$$|\mathbf{E} \cdot \mathbf{G}_{LS}(p)\mathbf{E}| > 0, \qquad \mathbf{E} \in \mathrm{Sym} \setminus \{\mathbf{0}\}, \qquad p \in \mathbb{C}^+,$$

is needed to ensure existence and uniqueness of the solution to the quasi-static problem (cf. §4.3). Then we need the further condition that $|\mathbf{E} \cdot \mathbf{G}(i\omega)\mathbf{E}| > 0$ for any nonzero $\mathbf{E}$ and real ω, namely

$$\text{(A.21)} \qquad |\mathbf{E} \cdot [\mathbf{G}_0 + \mathbf{G}'_F(\omega)]\mathbf{E}| > 0, \qquad \mathbf{E} \in \mathrm{Sym} \setminus \{\mathbf{0}\}, \qquad \omega \in \mathbb{R}.$$

The condition (A.21) means that the operator $\nabla \cdot [\mathbf{G}_0 + \mathbf{G}'_F(\omega)]\nabla$ is strongly elliptic in $H_0^1(\Omega)$.

By definition

$$\mathbf{G}_{LS}(0) = \mathbf{G}_\infty$$

and then $\mathbf{G}_{LS}(0) > 0$ is just the constitutive condition on the equilibrium elastic modulus. Now,

$$\Im[\mathbf{G}_0 + \mathbf{G}'_F(\omega)] = -\mathbf{G}'_s(\omega).$$

Then taking the thermodynamic condition on $\mathbf{G}'_s(\omega)$ in the stronger form (3.2.11), that is

$$\text{(A.22)} \qquad \omega\mathbf{G}'_s(\omega) < 0, \qquad \omega \in \mathbb{R},$$

provides the positive definiteness of $\Im[\mathbf{G}_0 + \mathbf{G}'_F(\omega)]$ in Sym for any $\omega \in \mathbb{R}^{++}$. Accordingly, the strong form (A.22) implies the validity of the inequality (A.21). In conclusion, $\mathbf{G}_\infty > 0$ and (A.22) are sufficient conditions for the validity of (A.20).

---

[42]Which follows from the thermodynamic conditions (A.8) and (A.9).

THEOREM A.4. *If* $\mathbf{G}$ *satisfies* (A.8), (A.22) *and* $\mathbf{G}_\infty > 0$, *then there exists* $\delta > 0$ *such that*

$$|\mathbf{E} \cdot [\mathbf{G}_0 + \mathbf{G}'_L(p)]\mathbf{E}| > \delta \mathbf{E} \cdot \mathbf{E}, \qquad \mathbf{E} \in \mathrm{Sym} \setminus \{\mathbf{0}\}, \qquad p \in \mathbb{C}^+. \tag{A.23}$$

Proof. For large values of p we have

$$\lim_{|p|\to\infty} \mathbf{G}_{LS}(p) = \mathbf{G}_0.$$

By (A.12) $\mathbf{G}_0 > \mathbf{G}_\infty$ and then there exist $\delta_1, \nu > 0$ such that

$$\Re \mathbf{E} \cdot [\mathbf{G}_0 + \mathbf{G}'_L(p)]\mathbf{E} > \delta_1 \mathbf{E} \cdot \mathbf{E} \quad \text{as } |p| > \nu.$$

Since $\mathbf{G}_0 + \mathbf{G}'_L(0) = \mathbf{G}_\infty > 0$, by (A.22) and the continuity of $\mathbf{G}_{LS}(p)$, there exist $\delta_2, \varepsilon > 0$ such that

$$|\mathbf{E} \cdot [\mathbf{G}_0 + \mathbf{G}'_L(p)]\mathbf{E}| > \delta_2 \mathbf{E} \cdot \mathbf{E} \quad \text{as } \Re p \in [0, \varepsilon).$$

Then by (A.10) there exists $\delta_3 > 0$ such that

$$|\mathbf{E} \cdot [\mathbf{G}_0 + \mathbf{G}'_L(p)]\mathbf{E}| > \delta_3 \mathbf{E} \cdot \mathbf{E} \quad \text{as } \varepsilon \le \Re p,\ |p| \le \nu.$$

Letting $\delta = \min\{\delta_1, \delta_2, \delta_3\}$ provides the desired result (A.23). □

References

[1] R.W. ATHERTON AND G.M. HOMSY, *On the existence and formulation of variational principles for nonlinear differential equations,* Stud. Appl. Math., 54 (1975), pp. 31–60.

[2] F. BAMPI AND A. MORRO, *The inverse problem of the calculus of variations applied to continuum physics,* J. Math. Phys., 23 (1982), pp. 2312–2321.

[3] G. BENTHIEN AND M.E. GURTIN, *A principle of minimum transformed energy in linear elastodynamics,* J. Appl. Math. Mech., 37 (1970), pp. 1147–1149.

[4] B. BERNSTEIN, E.A. KEARSLEY, AND L.J. ZAPAS, *A study of stress relaxation with finite strain,* Trans. Soc. Rheol., 7 (1963), pp. 391–410.

[5] D.R. BLAND, *The Theory of Linear Viscoelasticity,* Pergamon Press, Oxford, 1960.

[6] S. BOCHNER, *Lectures on Fourier Integrals,* Princeton University Press, Princeton, 1959.

[7] L. BOLTZMANN, *Zur Theorie der elastichen Nachwirkung,* Sitzber. Kaiserl. Akad. Wiss. Wien, Math.-Naturw. Kl., 70 (1874), pp. 275–300.

[8] R. BOWEN AND P.J. CHEN, *Thermodynamic restriction on the initial slope of the stress relaxation function,* Arch. Rational Mech. Anal., 51 (1973), pp. 278–284.

[9] G. CAPRIZ, *Sulla impostazione di problemi dinamici in viscoelasticità,* in Continui con Memoria, Accad. Naz. Lincei, Rome, 1990, pp. 25–33.

[10] G. CAPRIZ AND E.G. VIRGA, *Esempi di non-unicità in viscoelasticità lineare,* Atti Accad. Sci. Torino, 120 (1987), pp. 81–86.

[11] G. CAVIGLIA AND A. MORRO, *Noether-type conservation laws for perfect fluid motions,* J. Math. Phys., 28 (1987), pp. 1056–1060.

[12] ———, *Surface waves at a fluid-viscoelastic solid interface,* European J. Mech., A/Solids, 9 (1990), pp. 1–13.

[13] G. CAVIGLIA, A. MORRO, AND E. PAGANI, *Inhomogeneous waves in viscoelastic media,* Wave Motion, 12 (1990), pp. 143–159.

[14] P.J. CHEN, *On induced discontinuities behind shock waves in isotropic linear viscoelastic materials,* Nuovo Cimento B, 81 (1984), pp. 113–127.

[15] P.J. CHEN AND A. MORRO, *On induced discontinuities in a class of linear materials with internal state parameters,* Meccanica—J. Ital. Assoc. Theoret. Appl. Mech., 22 (1987), pp. 14–18.

[16] R.M. CHRISTENSEN, *Theory of Viscoelasticity,* Academic Press, New York, 1971.

[17] M. CIARLETTA AND M. PASQUINO, *Principio di minimo nella dinamica dei materiali viscoelastici,* Rend. Accad. Naz. Lincei, 69 (1980), pp. 147–152.

[18] M. CIARLETTA AND E. SCARPETTA, *Minimum problems in the dynamics of viscous fluids with memory,* Internat. J. Engrg. Sci., 27 (1989), pp. 1563–1567.

[19] B.D. COLEMAN, *Thermodynamics of materials with fading memory,* Arch. Rational Mech. Anal., 13 (1964), pp. 1–46.

[20] ———, *On thermodynamics, strain impulses, and viscoelasticity,* Arch. Rational Mech. Anal., 17 (1964), pp. 230–254.

[21] B.D. COLEMAN AND M.E. GURTIN, *On the growth and decay of one-dimensional acceleration waves,* Arch. Rational Mech. Anal., 19 (1965), pp. 239–265.

[22] ———, *Thermodynamics with internal state variables,* J. Chem. Phys., 47 (1967), pp. 597–613.

[23] B.D. COLEMAN AND V.J. MIZEL, *Norms and semi-groups in the theory of fading memory,* Arch. Rational Mech. Anal., 23 (1967), pp. 87–123.

[24] ———, *On the general theory of fading memory,* Arch. Rational Mech. Anal., 29 (1968), pp. 18–31.

[25] B.D. COLEMAN AND W. NOLL, *On certain steady flows of general fluids,* Arch. Rational Mech. Anal., 3 (1959), pp. 289–303.

[26] ———, *An approximation theorem for functionals, with applications in continuum mechanics,* Arch. Rational Mech. Anal., 6 (1960), pp. 355–370.

[27] ———, *Foundations of linear viscoelasticity,* Rev. Modern Phys., 33 (1961), pp. 239–249; *Errata,* 36 (1964), p. 1103.

[28] ———, *The thermodynamics of elastic materials with heat conduction and viscosity,* Arch. Rational Mech. Anal., 13 (1963), pp. 167–178.

[29] B.D. COLEMAN AND D.R. OWEN, *A mathematical foundation for thermodynamics,* Arch. Rational Mech. Anal., 54 (1974), pp. 1–104.

[30] A.D. CRAIK, *A note on the static stability of an elastoviscous fluid,* J. Fluid Mech., 33 (1968), pp. 33–38.

[31] C.F. CURTISS AND R.B. BIRD, *A kinetic theory for polymer melts. I. The equation for the single-link orientational distribution function,* J. Chem. Phys., 74 (1981), pp. 2016–2033.

[32] C.M. DAFERMOS, *On the existence and the asymptotic stability of solutions to the equations of linear thermoelasticity,* Arch. Rational Mech. Anal., 29 (1968), pp. 241–271.

[33] ———, *On abstract Volterra equations with applications to linear viscoelasticity,* J. Differential Equations, 7 (1970), pp. 554–569.

[34] ———, *Asymptotic stability in viscoelasticity,* Arch. Rational Mech. Anal., 37 (1970), pp. 297–308.

[35] ———, *Contraction semigroups and trend to equilibrium in continuum mechanics,* Lecture Notes in Math., 503 (1975), pp. 295–306.

[36] R. DATKO, *Extending a theorem of A. M. Liapunov to Hilbert space,* J. Math. Anal. Appl., 32 (1970), pp. 610–616.

[37] W.A. DAY, *Restrictions on relaxation functions in linear viscoelasticity,* Quart. J. Mech. Appl. Math., 24 (1971), pp. 487–497.

[38] ———, *The Thermodynamics of Simple Materials with Fading Memory,* Springer-Verlag, Berlin, 1972.

[39] ———, *Entropy and hidden variables in continuum thermodynamics,* Arch. Rational Mech. Anal., 72 (1976), pp. 367–389.

[40] ———, *An objection to using entropy as a primitive concept in continuum thermodynamics,* Acta Mech., 27 (1977), pp. 251–255.

[41] ———, *The thermodynamics of materials with memory,* in Materials with Memory, D. Graffi, ed., Liguori, Naples, 1979.

[42] W. DESCH AND R. GRIMMER, *Smoothing properties of linear Volterra integrodifferential equations,* SIAM J. Math. Anal., 20 (1989), pp. 116–132.

[43] E.H. DILL, *Simple materials with fading memory,* in Continuum Physics II, A.C. Eringen, ed., Academic, New York, 1975.

[44] G. DUVAUT AND J. LIONS, *Les inéquations en Mécanique et en Physique,* Dunod, Paris, 1972.

[45] M. FABRIZIO, *Entropy equation and absolute temperature for simple materials,* Boll. Un. Mat. Ital. Suppl. Fis. Mat., 2 (1983), pp. 123–136.

[46] ———, *Proprietà e restrizioni costitutive per fluidi viscosi con memoria,* Atti Sem. Mat. Fis. Univ. Modena, 37 (1989), pp. 429–446.

[47] ———, *An existence and uniqueness theorem in quasi-static viscoelasticity,* Quart. Appl. Math., 47 (1989), pp. 1–8.

[48] M. FABRIZIO AND C. GIORGI, *Sulla termodinamica dei materiali semplici,* Boll. Un. Mat. Ital. B, 5 (1986), pp. 464–474.

[49] M. FABRIZIO, C. GIORGI, AND A. MORRO, *Minimum principles, convexity, and thermodynamics in viscoelasticity,* Continuum Mech. Thermodyn., 1 (1989), pp. 197–211.

[50] M. FABRIZIO AND B. LAZZARI, *On the existence and the asymptotic stability of solutions for a linear viscoelastic solid system,* Arch. Rational Mech. Anal., to appear.

[51] ———, *On asymptotic stability for linear viscoelastic fluids,* to appear.

[52] M. FABRIZIO AND A. MORRO, *Thermodynamic restrictions on relaxation functions in linear viscoelasticity,* Mech. Res. Comm., 12 (1985), pp. 101–105.

[53] ———, *Viscoelastic relaxation functions compatible with thermodynamics,*

J. Elasticity, 19 (1988), pp. 63–75.

[54] ———, *Fading memory spaces and approximate cycles in linear viscoelasticity,* Rend. Sem. Mat. Univ. Padova, 82 (1989), pp. 239–255.

[55] ———, *Reversible processes in the thermodynamics of continuous media,* J. Nonequilib. Thermodyn., 16 (1991), pp. 1–12.

[56] G. FICHERA, *Existence theorems in elasticity,* in Encyclopedia of Physics, C. Truesdell, ed., Vol. VIa/2, Springer-Verlag, Heidelberg, 1972, pp. 347–389.

[57] ———, *Avere una memoria tenace crea gravi problemi,* Arch. Rational Mech. Anal., 70 (1979), pp. 101–112.

[58] ———, *Sul principio della memoria evanescente,* Rend. Sem. Mat. Univ. Padova, 68 (1982), pp. 245–259.

[59] ———, *Problemi Analitici Nuovi nella Fisica Matematica Classica,* Scuola Tipo-Lito Istituto Anselmi, Marigliano (Napoli), 1985.

[60] ———, *On linear viscoelasticity,* Mech. Res. Comm., 12 (1985), pp. 241–242.

[61] G.M.C. FISHER, *The decay of plane waves in the linear theory of viscoelasticity,* Brown University, Report NONR 562 (40)/2, Providence, RI, 1965.

[62] G.M.C. FISHER AND M.E. GURTIN, *Wave propagation in the linear theory of viscoelasticity,* Quart. Appl. Math., 23 (1965), pp. 257–263.

[63] C. GIORGI, *Alcune conseguenze delle restrizioni termodinamiche per mezzi viscoelastici lineari,* Quaderno n. 6/89, Università, Dipartimento di Matematica, Brescia, 1989.

[64] C. GIORGI AND B. LAZZARI, *Uniqueness and stability in linear viscoelasticity: some counterexamples,* Proceedings V Conference on Waves and Stability in Continuous Media, Sorrento, 1990.

[65] C. GIORGI AND A. MORRO, *Extremum principles for viscoelastic fluids,* Internat. J. Engrg. Sci., 29 (1991), pp. 807–817.

[66] D. GRAFFI, *Sui problemi dell'ereditarietà lineare,* Nuovo Cimento A, 5 (1928), pp. 53–71.

[67] ———, *Sull'espressione dell'energia libera nei materiali visco-elastici lineari,* Ann. Mat. Pura Appl., 98 (1974), pp. 273–279.

[68] ———, *Sull'espressione analitica di alcune grandezze termodinamiche nei materiali con memoria,* Rend. Sem. Mat. Univ. Padova, 68 (1982), pp. 17–29.

[69] ———, *On the fading memory,* Appl. Anal., 15 (1983), pp. 295–311.

[70] D. GRAFFI AND M. FABRIZIO, *Sulla nozione di stato per materiali viscoelastici di tipo rate,* Atti Accad. Naz. Lincei, 83 (1990), pp. 201–208.

[71] ———, *Non unicità dell'energia libera per i materiali viscoelastici,* Atti Accad. Naz. Lincei, 83 (1989), pp. 209–214.

[72] G. GRIOLI, *Continui con Memoria,* Accad. Naz. Lincei, Roma, 1990.

[73] H.D. GRUSCHKA AND F. WECKEN, *Gasdynamic Theory of Detonation,*

Gordon and Breach, New York, 1971.

[74] M.E. GURTIN, *Variational principles in the linear theory of viscoelasticity,* Arch. Rational Mech. Anal., 13 (1963), pp. 179–191.

[75] ———, *Variational principles for linear initial-value problems,* Quart. Appl. Math., 22 (1964), pp. 252–256.

[76] ———, *Variational principles for linear elastodynamics,* Arch. Rational Mech. Anal., 16 (1964), pp. 34–50.

[77] M.E. GURTIN AND I. HERRERA, *On dissipation inequalities and linear viscoelasticity,* Quart. Appl. Math., 23 (1965), pp. 235–245.

[78] M.E. GURTIN AND W.J. HRUSA, *On energies for nonlinear viscoelastic materials of single-integral type,* Quart. Appl. Math., 46 (1988), pp. 381–392.

[79] M.E. GURTIN AND E. STERNBERG, *On the linear theory of viscoelasticity,* Arch. Rational Mech. Anal., 11 (1962), pp. 291–356.

[80] I. HERRERA AND J. BIELAK, *A simplified version of Gurtin's variational principles,* Arch. Rational Mech. Anal., 53 (1973), pp. 131–149.

[81] I. HERRERA AND M.E. GURTIN, *A correspondence principle for viscoelastic wave propagation,* Quart. Appl. Math., 22 (1965), pp. 360–364.

[82] W.J. HRUSA AND M. RENARDY, *On wave propagation in linear viscoelasticity,* Quart. Appl. Math., 43 (1985), pp. 237–253.

[83] J. IGNACZAK, *A completeness problem for stress equations of motion in the linear theory of elasticity,* Arch. Mech., 15 (1963), pp. 225–234.

[84] E.F. INFANTE AND J.A. WALKER, *A stability investigation for an incompressible simple fluid with fading memory,* Arch. Rational Mech. Anal., 72 (1980), pp. 203–218.

[85] D.D. JOSEPH, *Slow motion and viscometric motion; stability and bifurcation of the rest state of a simple fluid,* Arch. Rational Mech. Anal., 56 (1974), pp. 99–157.

[86] ———, *Stability of Fluid Motions,* Springer-Verlag, Berlin, 1976.

[87] A. KAYE, *Non-Newtonian flow in incompressible fluids,* Tech Note 134, College of Aeronautics, Cranfield, England, 1962.

[88] H. KÖNIG AND J. MEIXNER, *Lineare systeme und lineare transformationen,* Math. Nachr., 19 (1958), pp. 256–322.

[89] E. KULEJEWSKA, *Functional methods of formulation of rheological constitutive potentials,* Arch. Mech., 36 (1984), pp. 67–76.

[90] H.M. LAUN, *Description of the non-linear shear behaviour of a low density polyethylene melt by means of an experimentally determined strain dependent memory function,* Rheol. Acta, 17 (1978), pp. 1–15.

[91] B. LAZZARI AND E. VUK, *Un teorema di esistenza e unicità per un problema dinamico in viscoelasticità lineare,* Atti Sem. Mat. Fis. Univ. Modena, 33 (1985), pp. 267–290.

[92] M.J. LEITMAN, *Variational principles in the linear dynamic theory of viscoelasticity,* Quart. Appl. Math., 24 (1966), pp. 37–46.

[93] M.J. LEITMAN AND G.M.C. FISHER, *The linear theory of viscoelasticity,* in Encyclopedia of Physics, Vol. VIa/3, C. Truesdell, ed., Springer-Verlag, Berlin, 1973.

[94] G. LUMER AND R.S. PHILLIPS, *Dissipative operators in a Banach space,* Pacific J. Math., 11 (1961), pp. 679–698.

[95] G.A. MAUGIN AND A. MORRO, *Viscoelastic materials with internal variables and dissipation functions,* Acta Phys. Hungar., 66 (1989), pp. 69–78.

[96] J.C. MAXWELL, *Constitution of bodies,* Sci. Papers, College Arts Sci. Univ. Tokyo, 2 (1877), pp. 616–624.

[97] V.J. MIZEL AND C.C. WANG, *A fading memory hypothesis which suffices for chain rules,* Arch. Rational Mech. Anal., 23 (1966), pp. 124–134.

[98] A. MORRO, *Temperature waves in rigid materials with memory,* Meccanica—J. Ital. Assoc. Theoret. Appl. Mech., 12 (1977), pp. 73–77.

[99] ———, *Negative semi-definiteness of the viscoelastic attenuation tensor via the Clausius-Duhem inequality,* Arch. Mech., 37 (1985), pp. 255–259.

[100] ———, *Topics in the Theory of Materials with Memory,* Mediterranean Press, Rende (Italy), 1990.

[101] A. MORRO AND M. FABRIZIO, *On uniqueness in linear viscoelasticity: a family of counterexamples,* Quart. Appl. Math., 45 (1987), pp. 263–268.

[102] A. MORRO AND M. VIANELLO, *Minimal and maximal free energy for materials with memory,* Boll. Un. Mat. Ital A, 4 (1990), pp. 45–55.

[103] A. NARAIN AND D.D. JOSEPH, *Linearized dynamics for step jumps of velocity and displacement of shearing flows of a simple fluid,* Rheol. Acta, 21 (1982), pp. 228–250.

[104] ———, *Classification of linear viscoelastic solids based on a failure criterion,* J. Elasticity, 14 (1984), pp. 19–26.

[105] W. NOLL, *A mathematical theory of the mechanical behavior of continuous media,* Arch. Rational Mech. Anal., 2 (1958), pp. 197–226.

[106] ———, *A new mathematical theory of simple materials,* Arch. Rational Mech. Anal., 48 (1972), pp. 1–50.

[107] A. PAZY, *On the applicability of Lyapunov's theorem in Hilbert space,* SIAM J. Math. Anal., 3 (1972), pp. 291–294.

[108] J. PRÜSS, *Positivity and regularity of hyperbolic Volterra equations in Banach spaces,* Math. Ann., 279 (1987), pp. 317–344.

[109] J.N. REDDY, *Variational principles for linear coupled dynamic theory of thermoviscoelasticity,* Internat. J. Engrg. Sci., 14 (1976), pp. 605–616.

[110] R. REISS, *Minimum principles for linear elastodynamics,* J. Elasticity, 8 (1978), pp. 35–45.

[111] M. RENARDY, *Some remarks on the propagation and non-propagation of discontinuities in linearly viscoelastic liquids,* Rheol. Acta, 21 (1982), pp. 251–254.

[112] M. RENARDY, W.J. HRUSA, AND J.A. NOHEL, *Mathematical Problems in Viscoelasticity,* Longmans Press, Essex, 1987.

[113] M. ROSEAU, *Asymptotic Wave Theory,* North- Holland, Amsterdam, 1976.

[114] P.E. ROUSE, *A theory of the linear viscoelastic properties of dilute solutions of coiling polymers,* J. Chem. Phys., 21 (1953), pp. 1271–1280.

[115] J.C. SAUT AND D.D. JOSEPH, *Fading memory,* Arch. Rational Mech. Anal., 81 (1983), pp. 53–95.

[116] M. SLEMROD, *A hereditary partial differential equation with applications in the theory of simple fluids,* Arch. Rational Mech. Anal., 62 (1976), pp. 303–321.

[117] ———, *An energy stability method for simple fluids,* Arch. Rational Mech. Anal., 68 (1978), pp. 1–18.

[118] R. TEMAM, *Navier-Stokes Equations,* North-Holland, Amsterdam, 1984.

[119] E. TONTI, *A general solution for the inverse problem of the calculus of variations,* Hadronic J., 5 (1982), pp. 1404–1450.

[120] F. TREVES, *Basic Linear Partial Differential Equations,* Academic Press, New York, 1975.

[121] C. TRUESDELL AND W. NOLL, *The non-linear field theories of mechanics,* in Encyclopedia of Physics, Vol. III/3, S. Flügge, ed., Springer-Verlag, Berlin, 1965.

[122] C. TRUESDELL AND R.A. TOUPIN, *The classical field theories,* in Encyclopedia of Physics, Vol. III/1, S. Flügge, ed., Springer-Verlag, Berlin, 1960.

[123] M.M. VAINBERG, *Variational Methods for the Study of Nonlinear Operators,* Holden-Day, San Francisco, 1964.

[124] E.G. VIRGA AND G. CAPRIZ, *Un teorema di unicità in viscoelasticità lineare,* Rend. Sem. Mat. Univ. Padova, 79 (1988), pp. 15–24.

[125] V. VOLTERRA, *Sulle equazioni integro-differenziali della teoria dell'elasticità,* Atti Reale Accad. Lincei, 18 (1909), pp. 295–301.

[126] ———, *Equazioni integro-differenziali della elasticità nel caso della isotropia,* Atti Reale Accad. Lincei, 18 (1909), pp. 577–586.

[127] ———, *Leçons sur les Fonctions des Lignes,* Gauthier-Villars, Paris, 1913.

[128] ———, *Sur la théorie mathématique des phénomènes héréditaires,* J. Math. Pures Appl., 7 (1928), pp. 249–298.

[129] ———, *Energia nei fenomeni ereditarii,* Acta Pontificia Acad. Scient., 4 (1940), pp. 115–128.

[130] C.C. WANG, *The principle of fading memory,* Arch. Rational Mech. Anal., 18 (1965), pp. 343–366.

[131] N.S. WILKES, *Thermodynamic restrictions on viscoelastic materials,* Quart. J. Mech. Appl. Math., 30 (1977), pp. 209–221.

[132] M.W. ZEMANSKY, *Heat and Thermodynamics,* McGraw-Hill, New York, 1957.

[133] B.H. ZIMM, *Dynamics of polymer molecules in dilute solutions: viscoelasticity, flow birefringence and dielectric loss,* J. Chem. Phys., 24 (1956), pp. 269–278.

Index